WORLD GUIDE TO
TROPICAL
DRIFT SEEDS
AND FRUITS

WORLD GUIDE TO TROPICAL DRIFT SEEDS AND FRUITS

Charles R. Gunn
and John V. Dennis

Illustrations by Pamela J. Paradine, F.L.S.

A DEMETER PRESS BOOK

Quadrangle / The New York Times Book Co.

Copyright © 1976 by Charles R. Gunn. All rights reserved, including the right to reproduce this book or portions thereof in any form. For information address: Quadrangle/The New York Times Book Co., 10 East 53 Street, New York, New York 10022. Manufactured in the United States of America. Published simultaneously in Canada by Fitzhenry & Whiteside, Ltd., Toronto.

Designed by Tere LoPrete

Library of Congress Cataloging in Publication Data

Gunn, Charles R 1927–
   World guide to tropical drift seeds and fruits.

   "A Demeter Press book."
   Bibliography: p.
   Includes index.
   1. Seeds—Tropics—Identification. 2. Seeds—Dispersal. 3. Beachcombing. 4. Fruit—Tropics—Identification. 5. Fruit—Dispersal. I. Dennis, John V., joint author. II. Title. III. Title: Tropical drift seeds and fruits.
   QK660.G84   1976   582'.03'34   75-36255
   ISBN 0–8129–0616–0

TO BETTY AND MARY ALICE
whose endless forbearance
has sustained us through
yet another venture

Contents

Preface	*ix*
1 *Introduction*	3
What are tropical seeds and fruits? Buoyancy. Plant dispersal. Economic aspects. Observations at sea. Samples from the sea bottom	
2 *History*	16
Northern Europe and United States. West Indies. Africa. Indian Ocean	
3 *Transport Currents and Collecting Beaches*	27
Northern Europe. Eastern United States. Gulf of Mexico. West Indies. Eastern South America. Africa. Indian Ocean. Australia and New Zealand. Eastern Asia. Pacific Ocean. Western New World	
4 *Collecting and Uses*	42
Suggestions for searching. Identification. A permanent collection. Tips for the foreign traveller. Enhanced beauty. Sea-bean jewelry. Instructions for growing	
5 *Systematic Descriptions and Illustrations*	58
Disseminule key	
CATALOG	67
Appendix	208
Generic list	
Family list	
Glossary	225
Bibliography	227
Index	233

Preface

Ocean beaches are depositories for as diverse a mixture of natural history objects and artifacts as can be found anywhere on dry land. It is an achievement for anyone to identify and understand even a small portion of the rubble that accumulates in the high tide zone. This book is prepared for those amateurs and professionals who find pleasure and excitement in one of the least studied of the beach treasures, stranded tropical seeds and fruits. These stranded seeds and fruits, collectively termed stranded disseminules or sea-beans, are not marine products. Rather, they are produced by terrestrial plants and transported by ocean currents from their place of origin to the beaches where they are stranded. Many drift disseminules never become stranded on a beach, meeting their fate on the bottom of the ocean because of loss of buoyancy.

The logistics of entering the ocean are simple for disseminules borne on plants that grow along the ocean. These disseminules fall onto the beach or into tidal pools or swamps and are carried to the ocean by the tide. The logistics are more complicated for disseminules of inland plants that grow along fresh water rivers. Their disseminules must be transported by the river to the ocean. These disseminules must be buoyant in fresh water, and the seeds should not germinate during this transport period, though many often do. Disseminules from littoral plants, plants which grow along the ocean, have a much better chance of being stranded in a suitable habitat than do those from inland plants.

While finding tropical disseminules on beaches within the tropics is interesting, the real excitement comes from the discovery of sea-beans on temperate beaches that may be hundreds or thousands of kilometers from the place where the parent plants grow. Some temperate beaches are noted for their masses of stranded disseminules, others may receive them occasionally, and most tem-

Preface

perate beaches receive none. The presence or absence of tropical disseminules on temperate beaches is a function of the juxtaposition of a strong tropical current. No range extensions result when tropical disseminules appear on temperate beaches. The cold climate precludes their establishment.

Few can resist the temptation to pick up tropical sea-beans. Often shiny and colorful, and foreign to the eye, they excite curiosity. Since early times, these seeds and fruits have captured the imagination of man, and have played a role in his history and everyday life. They have been used by ancient navigators as indicators of lands beyond the horizon, as "God's gifts" to be used in times of great need, and as objects of wonder or as ornaments.

This book is designed to serve the needs of both the amateur and the professional. Our data are gathered from a worldwide comprehensive collection of stranded disseminules and most of the relevant literature. Based on these collections and our field experience, we have assembled the first worldwide illustrated guide for tropical drift disseminules. In addition to describing and illustrating the frequently collected drift disseminules, we have evaluated the role that drifting has had on plant distribution, established buoyancy principles, circumscribed the two major centers of drift disseminules, and have tried to answer questions about the disseminules and the parent plants which may occur to the curious beachcomber.

We take full responsibility for the text. Many persons have contributed disseminules and knowledge, and we are greatly in their debt. Scores of collectors have generously furnished disseminules and local data and thereby contributed notably to the work. Several collectors have reviewed the typescript: Corinne E. Edwards (Colonel Sea Beans); Robert D. Mossman (Ole Jack Beans), whose collections inspired the authors; and Pat Parks, writer and journalist, who makes her home in the Florida Keys. Botanists who reviewed the typescript are Larry Fowler, Botany Department, University of South Florida, Dr. James A. Duke, Chief, Plant Taxonomy Laboratory, PGGI, ARS, U.S. Department of Agriculture, Dr. Richard Howard, Director, Arnold Arboretum of Harvard University, and Dr. C. Earle Smith, Jr., Plant Taxonomist, Department of Anthropology, University of Alabama. We deeply appreciate the Norwegian translations by Eeda Sissener Dennis, and the identifications of special taxa by plant taxonomists, whose names are recorded in the appropriate discussions. Other botanists and scientists who sup-

Preface

plied general information or identifications include Dr. C. Dennis Adams, University of the West Indies; Rosemary Angel, Royal Botanic Gardens; Stanley Kiem and George N. Avery, Fairchild Tropical Gardens; Dr. Jan Kohlmeyer, University of North Carolina; Dr. Elmar E. Leppik, Germplasm Resources Laboratory, U.S. Department of Agriculture; Ruth Mason, New Zealand Department of Scientific and Industrial Research; Dr. George Proctor, Institute of Jamaica; Dr. Gilbert Voss and Dr. Robert Work, University of Miami.

This book would not have been completed without the generous support of Dr. Edward S. Ayensu, Chairman, Department of Botany, Smithsonian Institution, Washington, D.C.

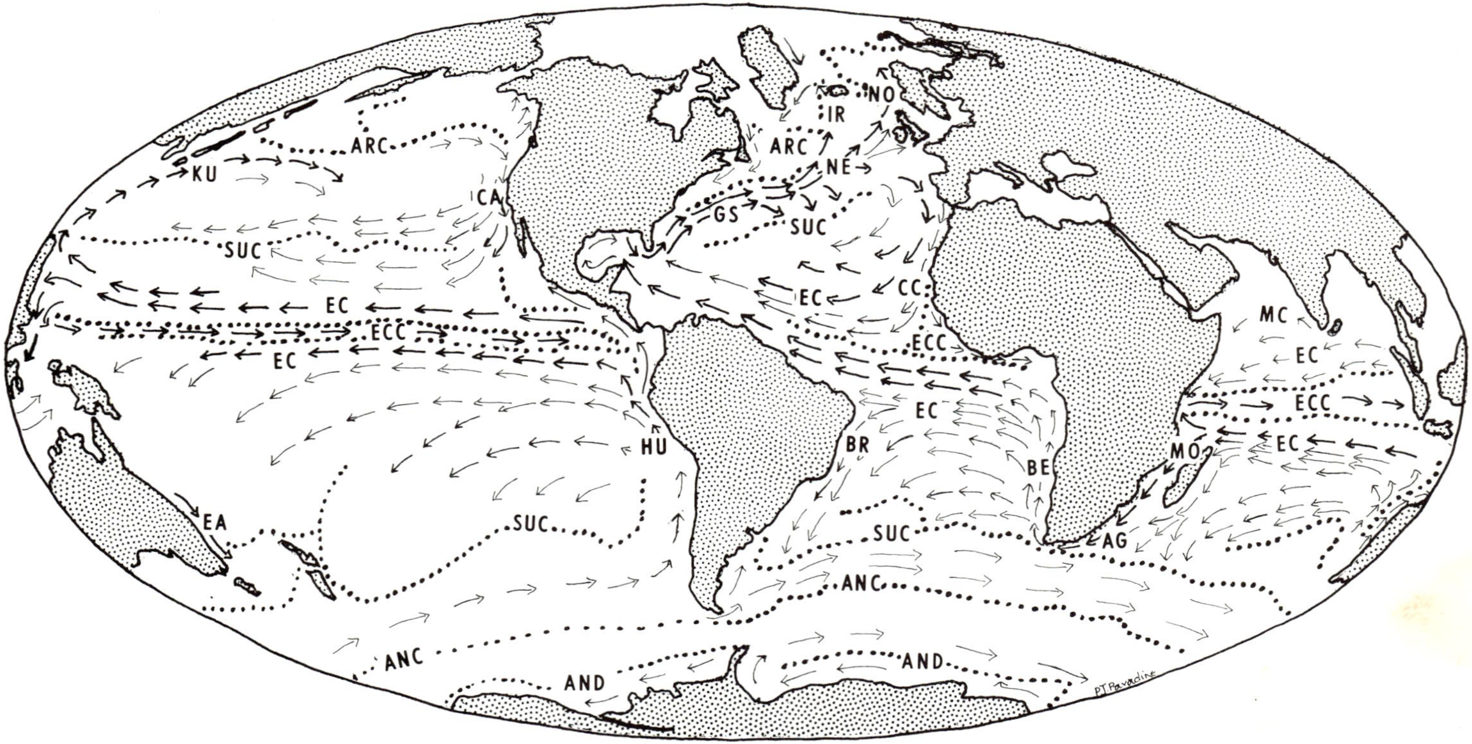
KU
ARC
CA
SUC
EC
ECC
EC
EA
SUC
ANC
AND
HU
GS
ARC
IR
NO
NE
SUC
EC
CC
ECC
EC
BR
BE
SUC
ANC
AND
AG
MC
EC
ECC
EC
MO
ANC

Currents named in the text are shown in this map. Darker arrows show the direction of currents with a velocity of more than 36 nautical miles per day, while lighter arrows show the direction of currents with a velocity of 36 or less nautical miles per day. Dots represent current boundaries. Arrow and dot locations are approximations.

Dots: ANC = Antarctic Convergence or AND = Antarctic Divergence
ARC = Arctic Convergence
SUC = Subtropical Convergence
Arrows: AG = Agulhas Current
BE = Benguela Current
BR = Brazil Current
CA = California Current
CC = Canary Current
EA = East Australia Current
EC = Equatorial Current
ECC = Equatorial Counter Current
GS = Gulf Stream
HU = Humboldt Current
IR = Irminger Current
KU = Kuroshio Current
MC = Monsoon Current
MO = Mozambique Current
NE = Northeast Atlantic Current
NO = Norwegian Current

WORLD GUIDE TO
TROPICAL
DRIFT SEEDS
AND FRUITS

I

Introduction

✥ *What are tropical drift seeds and fruits?*

Tropical seeds and fruits (disseminules or sea-beans) which drift to temperate beaches and become stranded have been collected by man for reasons that are as varied as there are collectors. For the romantic, the disseminules are messengers from exotic lands; for men of the sea, they represent victory over an ancient foe; for the superstitious, they represent gifts from the gods; and for the botanist, they are the end product of a plant dispersal mechanism. The tropical disseminules covered in this book have one common denominator. They all have the capacity to drift for at least one month in seawater. Viability is not a criterion. Some seeds were never alive, some lose their viability while drifting, while others are stranded in a viable condition.

Tropical drift disseminules may be true seeds, complete fruits, incomplete fruits, or seedlings. A true seed is a fertilized (rarely unfertilized) mature ovule that usually possesses an embryonic plant, stored food material (rarely missing), and a protective coat or coats. The embryo, or baby plant, is composed of one or more cotyledons, a plumule (embryonic bud), a hypocotyl (stem portion), and a radicle (rudimentary root). Application of the term "seed" is seldom restricted to this morphologically accurate definition. Rather, seed is usually used in a functional sense, viz., as a unit of dissemination, a disseminule. In this sense, the term seed embraces dry, one-seeded (rarely two- to several-seeded) fruits as well as true seeds. A fruit is a mature floral ovary that may contain one or more seeds and may include accessory floral parts. In this book the terms seeds and fruit are used in their correct botanical sense. Collectively seeds and fruits are labelled disseminules.

There are stranded objects which resemble seeds and fruits. The more common objects, such as pumice, cork, galls, thorns, and fungi, are depicted in Figures 2 and 3. None of these objects contain a plant embryo.

✥ Buoyancy

Most tropical disseminules do not float in either fresh water or seawater. We estimate that less than one percent of the tropical spermatophytes produce disseminules which drift in seawater for at least one month. The disseminules that drift do so because their specific gravity is less than that of seawater. The buoyancy principles that we have observed are placed in five groups which are listed below and illustrated in Figure 1. In enumerating these groups, we have considered the buoyancy classifications of Schimper (1891), Guppy (1906), and Muir (1937).

Group 1. Buoyancy due to cavity within the disseminule. Seed: Intercotyledonary cavity (*Entada* spp., *Ipomoea* spp., *Merremia* spp., *Mucuna fawcettii, M. sloanei, M. urens, Omphalea diandra*). Cavity formed because endosperm or embryo incompletely fills seed (*Aleurites* spp. and some individual seeds of *Caesalpina* spp.). Fruit: Cavities within fruit wall (*Juglans* spp., *Sacoglottis amazonica*). Cavity more or less central (*Canarium* spp., all genera of the Palmae).

Group 2. Buoyancy due to lightweight cotyledonary tissue (*Canavalia* spp., *Dioclea* spp., *Erythrina* spp.).

Group 3. Buoyancy due to a fibrous or corky coat, or a combination of both (*Cerbera* spp., *Hippomane mancinella, Terminalia* spp.).

Group 4. Buoyancy due to thinness of disseminule (*Avicennia germinans, Peltophorum inerme*).

Group 5. Buoyancy due to a combination of the above factors (*Barringtonia asiatica, Cocos nucifera, Grias cauliflora*).

The biological significance of the buoyancy phenomenon and of ocean currents as mechanisms of plant dispersal has often been inaccurately assessed. Two examples are cited: Darwin who dis-

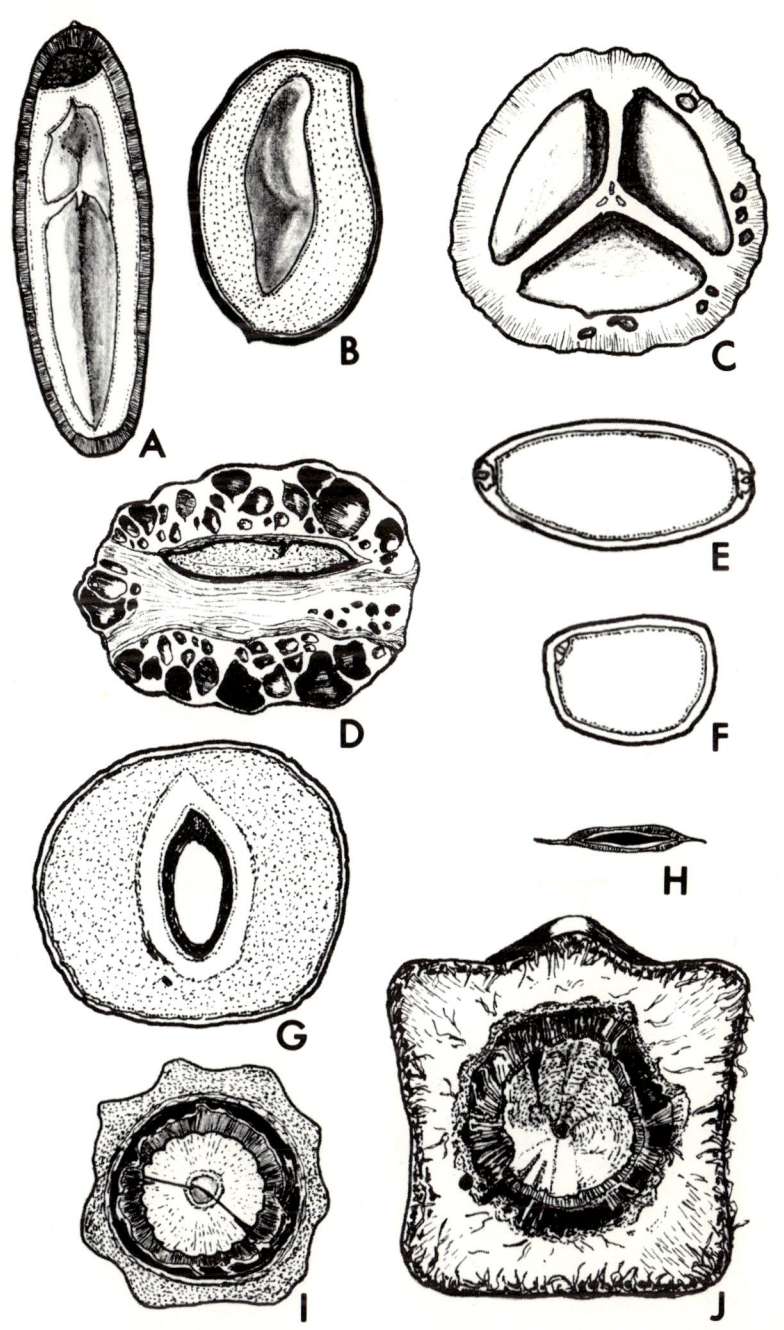

Figure 1. Buoyancy principles illustrated by cross sections or longitudinal section of drift disseminules. Group 1: A, *Entada phaseoloides*; B, *Omphalea diandra*; C, *Canarium decumanum*; D, *Sacoglottis amazonica*. Group 2: E, F, *Dioclea reflexa*, Group 3: G, *Cerbera manghas*. Group 4: H, *Peltophorum inerme*. Group 5: *Grias cauiflora*; J, *Barringtonia asiatica*.

Figure 2. Pumice and a cork float (lower left) may be found on beaches. Both float because of their light weight.

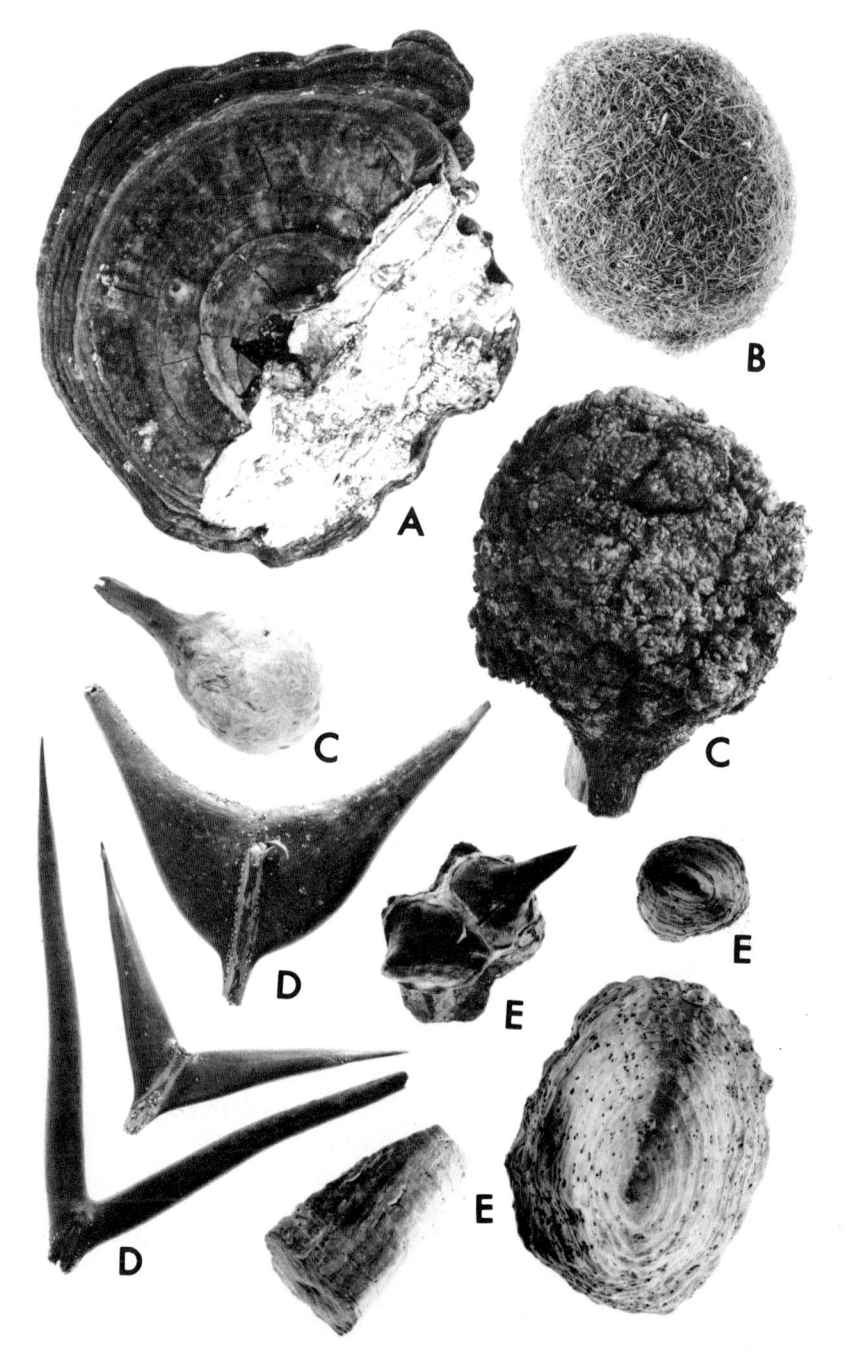

Figure 3. Stranded objects which resemble disseminules. A, fungus; B, ball of fibrous material collected along the coast of Morocco; C, galls; D, *Acacia* thorns; E, corky thorns of *Ceiba pentandra* (L.) Gaertner (X1).

(7)

counted the value, and Guppy who gave it too much value. Charles Darwin studied the role ocean currents played in the flora of Cocos-Keeling Islands in the Indian Ocean (Darwin, 1883). His conclusion, first published in 1836, was that most of the endemic vascular flora was derived from drift disseminules. This conclusion was contradicted by experiments which he conducted in England on seeds that were mostly garden species. He concluded that seeds in general did not float well in seawater, and furthermore they lost their viability quickly (Darwin, 1855, 1857). Like Candolle (1855), he saw little merit in ocean currents as a plant dispersal mechanism. Other scientists of the time arrived at similar conclusions.

Guppy (1906, 1917) and Ridley (1930), whose observations were derived from Guppy, championed ocean currents as a plant dispersal mechanism. In his zeal to overcome the nineteenth century lack of interest, Guppy made a strong case for the dispersal of certain species by ocean currents. His observations for the most part were accurate, and he supplied a much needed balance. Unfortunately he went too far, thus impairing the balance. Seeds of *Intsia bijuga* may be used to illustrate the point, though anyone of a number of disseminules could have been chosen. Buoyancy of *Intsia bijuga* seeds is governed by the weight of the cotyledonary tissue. There is no cavity like there is in *Entada, Caesalpinia,* and some *Mucuna* spp. seeds. Because the density of the cotyledonary tissue varies, some seeds sink, some barely float, and others are quite buoyant. This same variation is exhibited by the three named legume seeds which bear a poorly developed to well developed cavity. While Guppy regarded the cavity as nonadaptive, he regarded the buoyant cotyledonary tissue as adaptive. He attempted to correlate these adaptive buoyancy principles with where the seeds or fruits were produced. Returning to the *Intsia* example, Guppy theorized and tried to prove that these seeds seldom floated when produced by inland trees. On the other hand, seeds produced by littoral trees would usually float in seawater. Guppy regarded this as natural selection, a process by which buoyant disseminules were produced. We reject this theory as unsound, and we note that Guppy was never able to prove his theory in flotation tests, even though he made several attempts to do so. Our thoughts on the role of ocean currents as a plant dispersal mechanism are discussed and documented below.

Introduction

❧ Plant dispersal

The dissemination of terrestrial plant life by ocean currents is a haphazard process that is fraught with more failure than success. Even though viable disseminules may reach shorelines, this does not automatically mean establishment of the species. Moisture, soil conditions, competition, and foraging animals are some of the factors that must be overcome. Some drift disseminules tend to survive stranding, because their parents are weedy species which can survive the unstable conditions of the seashore (Carlquist, 1965).

Land crabs often damage seedlings on tropical beaches. Guppy (1890), when visiting the Cocos-Keeling Islands in the Indian Ocean, observed that viable disseminules were comparatively safe in the open sea, but that once they had stranded on the beach and had begun to germinate, the seedlings were quickly destroyed by robber crabs. The robber crab (*Birgus latro* Leach) poses a major threat to seedlings throughout its wide range in the Indo-Pacific region. English (1913) stated that two species of land crabs on Grand Cayman Island, West Indies are so destructive to seedlings that the self-establishment of drift species is highly unlikely.

We agree with Ridley (1930) that most drift species originated in tropical regions where there are many islands and not continental shores, such as those of South America and Africa. Ridley's view that the Malayan region was the major epicenter is sound. The islands of Indonesia have the most numerous drift flora of any region in the world. Currents passing this region carry disseminules both to the east into the vast expanses of the Pacific, and to the west across the Indian Ocean. The continent of Africa effectively blocks most movement westward beyond the Indian Ocean. A few species have reached the Atlantic and have achieved a pantropic distribution, because the Agulhas Current flows around the Cape of Good Hope.

The West Indies, the other major center, has fewer species than Indonesia. These islands are the center from which a number of New World disseminules have radiated. Movement to and from the west may be assumed to have taken place during past geological times when the Atlantic and Pacific were not separated by a continuous land barrier in the region of the Isthmus of Panama. Some species reached the Pacific coast of the New World before the Pleiocene age, when the two oceans were joined. Of the 39 drift species whose disseminules are found in the drift of San Jose Island in the Gulf of Panama, 29 are common to both sides of the Isthmus of Panama (Johnston, 1949).

Oceans may be barriers as well as avenues of transport. This is seen not only in the South Atlantic, where there is little possibility of eastward movement because of absence of currents, but also in the Pacific where another limiting factor exists. Wide expanses of the eastern Pacific are virtually without islands. This appears to have greatly curtailed the spread of drift species in this part of the world. Johnston (1949), in commenting upon this phenomenon, observed that a New World mangrove (*Rhizophora samoensis*) is the only New World drift species that has succeeded in reaching Polynesia. Ridley (1930) listed a number of others, including *Dioclea reflexa*, *Mucuna urens*, *Sophora tomentosa*, and *Rhizophora mangle*, which he suggested were of New World origin and had crossed the eastern Pacific to Polynesia. We consider these species to be of Old World origin, and therefore disagree with Ridley. Guppy (1906) noted that there are progressively fewer drift species of Old World origin established in the island groups of the Pacific as one travels eastward. He points to about 65 such species in the Fiji Islands, about 40 in Tahiti, about 16 in Hawaii, and a few in islands as far eastward as Pitcairn. None of the Old World drift species have crossed the remaining expanse of the Pacific to reach the shores of the New World. This may be due to the absence of transport currents. The Equatorial Counter Current is an unlikely conveyer of drift material. Although maps show a well-marked counter current moving across the Pacific in the vicinity of the equator, this current, in the words of Heyerdahl (1968), is "nothing but an interrupted series of upwellings" and, "is of scant use to trans-Pacific voyagers."

In view of the many obstacles, it is still not surprising that drift disseminules have been distributed by ocean currents, because of the vast numbers of disseminules which are produced each year. Among the likely carriers of small seeds are pieces of driftwood, pumice (Fig. 1), and certain corals which are buoyant enough to float in seawater. Stems, rhizomes, and other vegetative parts may be transported by ocean currents. These may take root on becoming stranded, even though there has been a long exposure to seawater.

Any object that is floating on the surface of the ocean is at the

Introduction

mercy of wind, waves and, if in a current, the direction that this current is moving. This means that ocean transport is unreliable. Many currents which originate in the tropics turn north or south, and therefore tropical disseminules are carried to inhospitable regions where there can be no expectation of establishment. Even within the tropics, some coastlines receive rich desposits of drift disseminules, while other coastlines receive little or nothing from distant sources. This is a matter that is not only determined by the presence or absence of an offshore tropical current but by other factors, including prevailing wind direction and the topography of the shoreline. The result is that drift species are unevenly distributed. Some coastlines in the tropics have numerous species: other coastlines have only a modest number, or none.

Economic aspects

Ocean currents have aided the dispersal of a number of species that are useful to man. Mangroves, for example, assist in building new land. The coconut, whose dispersal has been aided to some extent by ocean currents, is one of the most important trees known to man. Some sea-beans which take a high polish are useful as ornaments and have other uses.

Little is known, however, of the economic implications of the movement of tropical drift, including drift disseminules, by way of ocean currents into temperate portions of the oceans. This might seem to be one of the more harmless manifestations of nature. However, the discovery of tropical marine fungi on red mangrove seedlings that had stranded on Bogue Banks near Atlantic Beach, North Carolina (Kohlmeyer, 1968) indicates that this might not be true. Until recently marine fungi have been a little-studied group. Certain tropical species are wood-rotting agents without whose assistance shipworms and teredos could not damage pilings and other wooden structures in seawater (Kohlmeyer and Kohlmeyer, 1971). Tropical marine fungi become attached to floating objects and are thereby transported into more northern waters. The Kohlmeyers recognize a potential hazard when they suggest that tropical marine fungi that are regularly transported to northern waters may in time become established in tidal rivers and estuaries near electric power plants with thermal discharges.

Although other mangrove species and the coconut have been implicated as carriers of tropical marine fungi, it is the large, abundant seedlings of the red mangrove that are suspected of playing the most important role. Seedlings of the red mangrove may float for a year or more in seawater. Hence, there is a potential danger that tropical marine fungi could be carried to the North Atlantic. Red mangrove seedlings have not yet been reported from beaches of western Europe.

Observations at sea

Thousands of drift disseminules strand along a few miles of beach. Yet observers, who in recent years have watched for the disseminules in the ocean, report negative results. This has been true even when observers have watched from a small open boat. Plankton sampling has been conducted in various places. Yet no reports have reached us of floating disseminules being retrieved.

Smaller disseminules might easily escape notice by an observer on shipboard, but a number, including the coconut, calabash, and box fruit, are large and tend to float fairly high in the water. Along the east coast of Florida we have seen the coconut and also red mangrove seedlings floating within a few meters of the shore. Sometimes whole flotillas of red mangrove seedlings pass by, each seedling floating vertically, as a rule, and bobbing up and down in the water as though animated by a form of propulsion.

In earlier literature pertaining to tropical drift disseminules, there are occasional references to sightings at sea. Most such observations were made close to land, and presumably, in many cases, from small boats or sailing craft. We have searched in vain for any mention of numbers seen and details of abundance and depth.

Bates (1863) observed many sea-coconut fruits floating at sea some 650 km from the mouth of the Amazon. Hemsley (1885), reporting upon observations made by Guppy, stated that solitary box fruits were seen from shipboard some 250 km at sea, south of the Solomon Islands. Guppy (1890), in discussing the drift on beaches of the Cocos-Keeling Islands in the Indian Ocean, states that a voyager nearing the Islands can observe various drift seeds and fruits floating on the surface of the ocean. Perhaps the most interesting observation was made by Guppy (1906) who, in telling of the

Introduction

drift along the coast of Ecuador, stated that coralbean (*Erythrina*) seeds were abundant in both floating and stranded drift. He observed seeds many kilometers out to sea beyond the mouths of major rivers. Because coralbean seeds are only 1 to 1.8 cm long, they must have been present in great numbers or have been seeds of bright colored species to be seen drifting in the ocean.

✑ *Samples from the sea bottom*

Of the countless number of tropical drift disseminules which enter the sea, most presumably sink to the bottom. Guppy (1906) spoke of dredging operations in the harbor at Honolulu, Hawaii that brought up quantities of old candlenut fruits. Another discovery was the dredging of a mango endocarp off Bogue Banks, North Carolina (Gunn and Dennis, 1972b).

We learned that the Rosenstiel School of Marine and Atmospheric Science, University of Miami, had jars containing bottom samples obtained through deep sea dredging operations in parts of the Caribbean, the Gulf of Panama, and neighboring waters, as well as samples from the Gulf of Guinea, Africa. We searched these jars in order to find out if there were sea-beans present.

Some 20 species of tropical drift disseminules were found. These were in gallon jars from widely scattered dredging locations. The number of disseminules in jars containing disseminules varied from one to fifteen. Among the disseminules obtained from dredging in New World waters were those of antidote vine (*Fevillea cordifolia*), bloodwood (*Pterocarpus officinalis*), coconut (*Cocos nucifera*), country-almond (*Terminalia catappa*), hog-plum (*Spondias mombin*), mango (*Mangifera indica*), red mangrove (*Rhizophora mangle*), sea-coconut (*Manicaria saccifera*), and white mangrove (*Avicennia germinans*). The assortment was not unlike that which could be found on beaches in the same parts of the world. Pieces of pumice were found along with the disseminules, and also quantities of vegetable debris that included leaves, coconut husks, tropical thorns, pieces of wood, bamboo, and stalks of sugar cane.

Samples from the Puerto Rico Trench, an ocean depth that lies in the Atlantic to the east of Puerto Rico, not only contained on the average more species, but the specimens were unusually well preserved (Fig. 4). The great depths at which the samples were re-

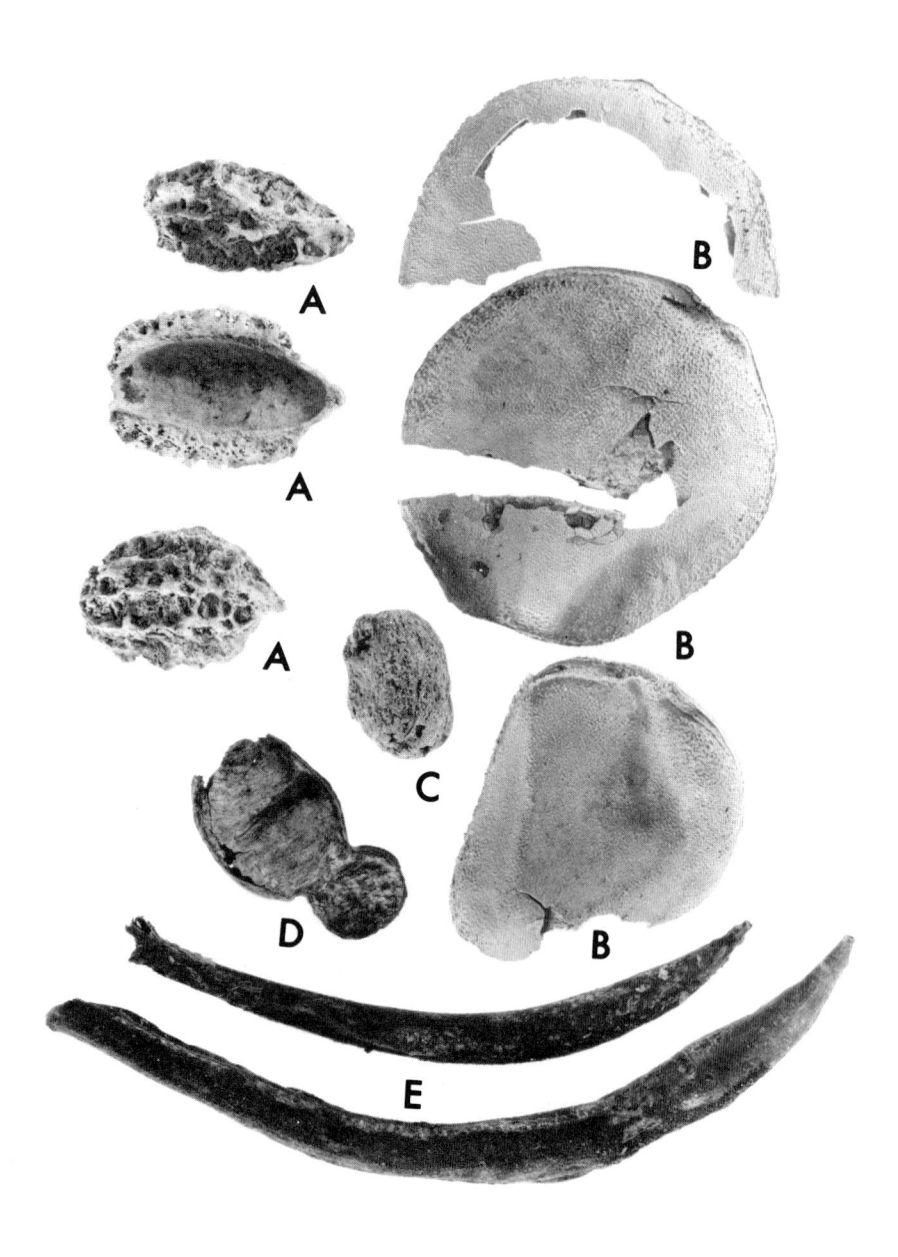

Figure 4. Disseminules dredged from the Puerto Rico Trench in jar numbered 1168 at the Rosensteil School of Marine and Atmospheric Science, Miami. A, *Terminalia catappa* (country-almond); B, *Fevillea cordifolia* (antidote vine); C, *Spondias mombin* (hog-plum); D, partial legume pod; E, *Rhizophora mangle* (red mangrove) (X1).

(14)

Introduction

trieved, up to 7500 m, may have been a factor in the excellent state of preservation that was noted, because fewer decay organisms were present.

Legume seeds were the only important drift element missing from these bottom samples. We were surprised, because hard-coated seeds of *Mucuna, Entada, Caesalpinia,* and *Dioclea* are numerous on beaches. Our hypothesis is that the legume seeds that drift are so buoyant and impervious to seawater that they rarely sink.

2

History

Tropical drift disseminules, popularly known as sea-beans, have had a long and colorful history. It is impossible to record all of the facts, superstitions, and lore which have been recorded about sea-beans. We have selected some of the more interesting stories from various parts of the world. Perhaps the most significant one is told about Christopher Columbus. According to tradition, the sea heart (*Entada gigas*) provided inspiration to Columbus and led him to set forth in search of lands to the West. The sea heart is called *favas de Colom*, or Columbus-bean, by the inhabitants of Porto Santo, Azores. During his first voyage, when he landed on the coast of northeastern Cuba, Columbus recorded seeing "a large nut of the kind belonging to India" (Morison, 1942). Some writers have used this vague reference as support for claims that the coconut had a much wider New World distribution at the time of its discovery than is generally acknowledged. We believe that Columbus did not see the true coconut (*Cocos nucifera*) (Gunn and Dennis, 1972).

Northern Europe and United States

The Gulf Stream brings tropical disseminules to Northern European beaches from the Caribbean Sea region. The most frequent stranded sea-beans are the sea heart (*Entada gigas*), true sea-bean (*Mucuna sloanei* and *M. urens*), sea purse (*Dioclea reflexa*), gray nickernut (*Caesalpinia bonduc*), and Mary's-bean (*Merremia dis-*

coidesperma). Most, if not all, of the following discussions involve one or more of these drift seeds (Fig. 5).

Pena and L'Obel (1570) may have been the first to record the occurrence of stranded tropical seeds on a temperate beach. While they did not describe the seeds, they referred to them as "beans." They did note that the collector, Dame Catherine Killigrew, observed that there were no shipwrecks in the sea off the Cornwall, England coast where she had made the collections. In light of later controversy, it is worth noting that they were surprisingly accurate in their thoughts, regarding the vegetable origin of the "beans" and their probable source. They proposed that the seeds had drifted from the New World by "favorable southerly or westerly winds." This is as close to the truth as anyone could possibly come at a time when the Gulf Stream was unknown. The prevailing view among coastal inhabitants was that these sea-beans grew on underwater trees or tangles.

Among the Scandinavians there was an even more primitive tradition. The sea-beans were regarded as stones and given such descriptive names as confinement stone, fat kidney, bent stone, and worm stone. A Norwegian Bishop (Gunnerus, 1765) recorded some of the Norse uses of the sea heart. A woman in childbirth could seek relief from pain by drinking a strong brew or ale from a cup made from the sea heart seed coat. The sea heart embryo was also used as a purgative and as a medicine for cattle. The taste was bitter, like quinine, and was more pronounced in some seeds than in other seeds. Sea hearts were commonly used in Norway and other parts of Europe as snuffboxes. The seeds were cut in half, the embryo removed, and the exterior polished. Gunnerus mentioned that people of modest means were content with unadorned snuffboxes, while those of wealth sported boxes that were hinged and richly adorned with silver. The true sea-bean (*Mucuna* spp.) was also used in making snuffboxes.

In England the sea heart was used for a teething ring. Seeds were given as good luck pieces to children who were likely to go down to sea, because the seeds had been washed ashore in a sound condition. Martin (1703) recorded the lore of the Hebrides Islanders. The gray nickernut (*Caesalpinia bonduc*) was worn as an amulet and was useful in warding off the Evil Eye. The seeds were supposed to turn black when harm was intended to the wearer. The magical

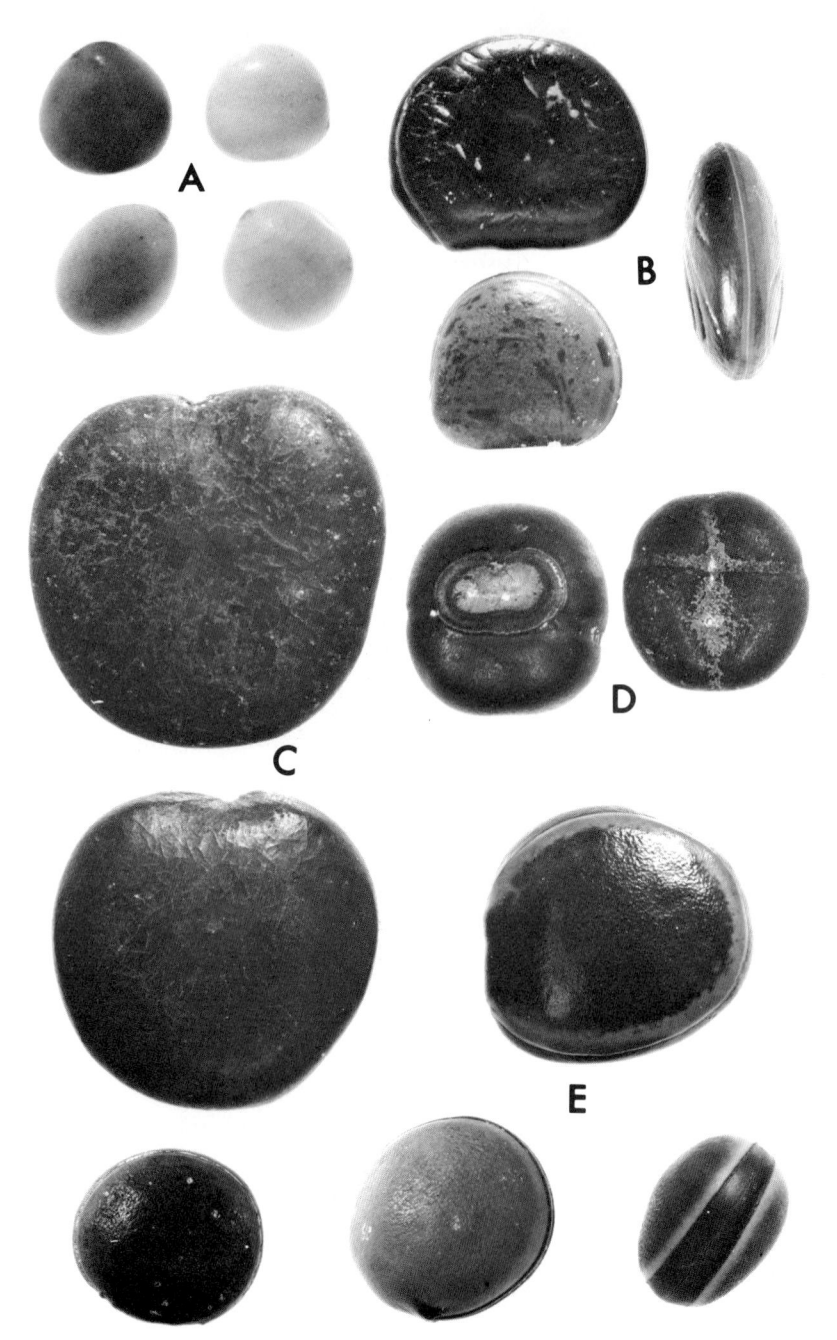

Figure 5. Tropical drift seeds stranded on the western coast of Ireland. A, 4 seeds of *Caesalpinia bonduc* (gray nickernut); B, 3 seeds of *Dioclea reflexa* (sea purse); C, 2 seeds of *Entada gigas* (sea heart); D, 2 seeds of *Merremia discoidesperma* (Mary's-bean); E, 4 seeds of *Mucuna* spp. (true sea-bean) (X1).

powers of this seed, known in the Hebrides as the white Indian nut, extended to clearing up blood in cow's milk. When a nickernut was placed in a pail of milk, the milk was supposed to clear up almost immediately. The powdered embryo of the seed was said to be a cure for dysentery when taken in boiled milk.

Perhaps the most precious stranded seed was, and still is to many, Mary's-bean (*Merremia discoidesperma*). This seed, also known as crucifixion-bean, has a rich history and was mentioned in early accounts from Ireland, the Hebrides, the Orkneys, and Shetland Islands. One distinctive feature of these seeds is the cross stamped on one surface. To pious people who found these stranded seeds, the cross gave the seeds special meaning. The seeds had obviously survived the ocean and would now extend their protection to anyone lucky enough to own one. In the Hebrides a woman in labor was assured an easy delivery, if at the proper time, she clenched a seed in her hand. Seeds were handed down from mother to daughter as treasured keepsakes.

Western Ireland has had a long history pertaining to sea-beans (Colgan, 1919). However, the amount of folklore is meager, because no one recorded it. Colgan noted that "no doubt a fund of folklore still lingers round these mysterious sea-waifs in the minds of the wise women of our western coasts. Such lore, however, is not to be extracted without patient and skillful manipulation." At one time prudent residents along the western coast placed a sea-bean under the pillow at night as a charm against the nocturnal visits of fairies.

Europeans have sought various explanations about the origin and transport mechanism of stranded sea-beans. As previously noted, most rural people thought that the seeds came from underwater plants. The common names Molucca-bean or crospunk for these sea-beans were prominent in the vocabulary of the islanders along the Scottish coast by the early seventeenth century (see crospunk in Warrack, 1965). Some authors, including Sloane (1725), thought that the name Molucca-bean arose from the belief that the seeds came from the Moluccas or Spice Islands of the East Indies. The islanders were also said to give credence to the existence of a "northeast passage" which transported the seeds from the Pacific to the Atlantic Ocean. From our examination of the literature, we can find no evidence to support such sophisticated theories. It seems likely that the imagination of the islanders had been kindled by stories of the Spice Islands, and that they had simply attached a romantic name to the sea-beans. An obscure origin gave the sea-beans an aura of mystery. This, coupled with their natural beauty and indurate surface, was enough to assure their popularity.

Sir Hans Sloane, founder of the British Museum, was one of the first scientists to identify a potential source of the sea-beans. Based on his study of the Jamaican flora and the disseminules he saw from Scottish beaches, he concluded that these disseminules had their origin in the West Indies. He thought that the seeds were carried northward by the current from the Gulf of Florida (now the Gulf of Mexico) and then drifted eastward across the Atlantic under the influence of westerly winds (Sloane, 1696). The current he alluded to was discovered by Ponce de Leon in 1513 in the vicinity of Cape Canaveral, Florida. The current is now called the Gulf Stream. Sloane's scientific note was followed by a description of life in the Orkneys in which four sea-beans were clearly illustrated (Wallace, 1693 and 1700).

The Danish historian Lucas Debes, in his notes on the Faeroe Islands, noted that a "very knowing man" assured him that the sea-beans came from the West Indies, and that they were brought by "the stream." A chart prepared by Kircher in 1678 showed the current we now know as the Gulf Stream approaching Europe, but the distribution of this current map appeared to be very limited. One of the first widely distributed and reasonably accurate maps of the Gulf Stream was made by Benjamin Franklin in 1770 (Griswold, 1951). The Gulf Stream in this chart is not named, and it is shown going past the Grand Banks of Newfoundland. By the beginning of the nineteenth century, the Gulf Stream concept was generally accepted. Its eastward flow after leaving the New World was still open to question. The work of Gumprecht (1854) finally established the true scope of the Gulf Stream System. It was his carefully documented records of the occurrences of stranded tropical drift disseminules and other tropical debris that clearly established the fact that the Gulf Stream System did reach northern Norway. Although the Gulf Stream transports tropical drift disseminules to New World beaches from Cape Cod, Massachusetts to the Florida Keys, we have found few historical notes.

History

⋄§ West Indies

As in northern Europe, the sea-beans which have a prominent history are the sea heart, true sea-bean, sea purse, gray nickernut, and Mary's-bean. Some of the uses were mentioned by Sloane (1707). He gathered the information while serving as physician to the governor of Jamaica from 1687 to 1689. He noted that the sea heart, true sea-bean, and gray nickernut were used as coat buttons, snuffboxes, medicines, and toys. Gray nickernut seeds are still used as marbles. Another name for marble is nicker. All of the mentioned seeds are still used in making seed jewelry. One use that is close to the heart of small boys is to use any of these seeds as burning-beans. By rubbing these seeds vigorously on cloth, the surface of the seed becomes quite hot. The hot seed when touched to someone's skin (especially an unsuspecting girl) gives the victim a start. While all of these seeds may be called burning-beans, the name is usually applied to the true sea-bean and the gray nickernut. The uses of Mary's-bean by Christians have been noted. Other interesting notes about Mary's-bean seeds may be found in Gunn (1976) and Williams (1973).

⋄§ Africa

East Africa is the source of the South Atlantic Equatorial Current. This transport current is the only tropical current which unites the Old World with the New World. Pantropic drift species like the sea heart, true sea-bean, sea purse, and gray nickernut were transported from East Africa to the Caribbean region and adjacent continental coast. Disseminules whose origins are in the New World cannot drift to Africa, because the South Atlantic Counter Current is an ineffective transport current.

The uses of the seeds listed above are essentially the same as their uses in the West Indies. This is not surprising when one considers that another uniting factor between Africa and the West Indies was slavery. Negroes brought to the West Indies as slaves also brought their traditional uses of these seeds, as medicines, toys, and counters in games. Muir (1937) described the uses he noted in East, West, and South Africa. One of the most common uses was to ward off sickness. The true sea-bean, with a thong attached, is carried as an amulet or charm to protect the owner from sickness. In some parts of west Africa, seeds of the true sea-bean and the sea purse were regarded as coming from one plant. Seeds of the true sea-bean, which are somewhat duller, were designated the male.

Indian Ocean

If any of the world's drift disseminules is entitled to a major place in romanticism and folklore, it is the coco-de-mer (*Lodoicea maldivica*). The nut was held in high esteem, because it was believed to possess medicinal properties, and because its endosperm was believed to be an effective antidote against poison. Owners of nuts for sale would also tout the value of the nuts as aphrodisiacs, no doubt trying to increase their price. Until the discovery of the parent plants on Praslin and Curieuse Island in the Seychelles by Mahé de la Bourdonnais in 1743, the only source of the nuts was the drifting or stranded nut.

The most productive beaches for the coco-de-mer were those on the Maldive Islands, southwest of India. It was a policy among the local rulers to commandeer any stranded coco-de-mer nuts and to sell them to rich Europeans for as much money as they could. During this period a single coco-de-mer was worth the price of a loaded ship. It was no wonder that death was the punishment for anyone who did not promptly surrender a stranded coco-de-mer.

So long as the coco-de-mer was in limited supply, claims concerning its powers as a cure-all became ever more grandiose. It was supposed to be a cure for all inflammations of the body and a preventative against colic, apoplexy, epilepsy, and paralysis. Even the shell was thought to have special antiseptic properties and was fashioned into vessels to keep water fresh and free from contamination (Fig. 6). The sick became immune to other diseases if they drank water that had been kept in the shell for some time, and to which had been added a piece of the endosperm. Dishes, cups, and containers were made from the shell, and there is still a brisk trade in such objects in the Seychelles (Figs. 6 and 7). After the islands were discovered in 1743, the value of the nuts fell sharply and much less was heard of their very exceptional properties.

Early stories concerning the origin of the nuts were highly imag-

Figure 6. An excellent example of an ornate coco-de-mer (Clusius, 1605).

Figure 7. Coco-de-mer. A, Dr. Gunn holding a longitudinal section of a coco-de-mer in his right hand and an entire endocarp in his left. B and C, polished endocarps bearing some carvings and hinged; courtesy of Mr. & Mrs. Wendell B. Coote.

History

inative. According to Rumphius (Blatter, 1926), Malay and Chinese sailors regarded the parent plant as an underwater tree similar to a coconut which was below the surface in placid bays along the coast of Sumatera (Sumatra). If anyone attempted to dive to the tree, the shadowy outline disappeared instantly. In another version of the story, a huge bird or griffin was said to live in branches rising above the water. The trees with their birds were found off the coast of Djawa (Java). Nightly the birds would sally forth to prey upon tigers, elephants, and rhinoceroses, the flesh of which was carried back to the nest. Ships were attracted by the waves which surrounded the trees, and mariners at such times were said to be preyed upon by the savage birds.

These accounts, however far-fetched, were in accord with the prevailing view that the tree producing the nuts either grew partially submerged or wholly underneath the surface of the sea. When the coco-de-mer trees were discovered, they were usually found in rocky uplands. They were not especially adapted to seaside conditions. The species never spread from its two home islands, because the seawater killed the nuts before they could reach another beach. Even today viable nuts are hard to obtain.

Perhaps the most unusual true story about any drift disseminule is one concerning the coco-de-mer, General Gordon, and the Garden of Eden. General Gordon was the archetype of a nineteenth century British soldier-administrator. As a leader of the time, he was well known for his deeds, piety, bravery, resolution, unselfishness, and hero's death at Khartoum three years after the discovery of his Garden of Eden. General Gordon was sent by the British War Office in 1882 to survey the Seychelles as a potential British base. While he decided that building the base was too costly, he did discover on Praslin the *Valle de Mai*—the Valley of the Giants. He became convinced that this most beautiful valley was the biblical Garden of Eden and that the coco-de-mer trees were the Tree of Knowledge of Good and Evil. These trees are the giants which are referred to in the phrase, Valley of the Giants. Several reasons have been offered to explain why General Gordon saw this valley as the Garden of Eden and the coco-de-mer as the "apple tree." Two patent explanations are the exceptional beauty of the valley and the undeniable resemblance of the coco-de-mer nut, its "form evocative," to certain portions of a woman's anatomy. From notes and drawings filed at the Royal Botanic Gardens Library, Kew, England, there can be absolutely no doubt that General Gordon had found his Garden of Eden. For many who regarded him as the man who saved China, it is fitting that he found his paradise on earth before dying in his single-handed defense of Khartoum.

3
Transport Currents and Collecting Beaches

Those who are used to collecting sea shells will have no difficulty in knowing how to search for sea-beans. However, there are important geographical differences in where to look. Except for a few pelagic species, sea shells are a product of the ocean bottom near the beach where they are stranded, and a number of them may be found on any beach. Sea-beans may be expected on most tropical beaches, and ocean currents may carry them from the tropics to some temperate beaches, rarely to arctic beaches. To find sea-beans outside the tropics, a transport current must be available. Otherwise there is no use looking for them. Currents are shown in the map on page *xii*.

Northern Europe

It is a great thrill to find a sea-bean on a beach that is thousands of kilometers from the tropics where the parent plant grew. Unexpected collecting opportunities exist on many mainland and island beaches in more northern reaches of the Atlantic. The presence of tropical debris on beaches from Greenland and Iceland to Norway and the British Isles may be explained by referring to the map showing the Gulf Stream System. This mighty current system, with its origin in the Gulf of Mexico, is one of the world's great carriers of tropical debris. The several currents that contribute to its flow may bring tropical debris from the Caribbean region, adjacent Central America, and northern South America. Some debris may come by way of the South Atlantic Equatorial Current from as far away as the west coast of west Africa.

In the vicinity of the Grand Banks of Newfoundland, the Gulf Stream divides. A western arm known as the Irminger Current flows toward Iceland and brings to that island's southern shores tropical driftwood and sea-beans. Some of this debris is carried by the Irminger Current to the southern shores of Greenland. The sea heart is said to be well known to residents of Danish settlements in the southern part of Greenland.

A northeastern arm, the Northeast Atlantic Current, flows toward western Europe, becoming the Norway Current as it passes the Faeroe and Shetland Islands. This current is roughly less than a third the volume of the Gulf Stream; yet its influence in ameliorating the climatic conditions of northern Europe is significant. The waters of this current enter the Barents Sea and bring sea-beans and other tropical drift material to such northern outposts as Nova Zembla and Spitzbergen. A sea heart from the northern coast and a gray nickernut from the west coast of Spitzbergen represent the northernmost records for any of the tropical drift disseminules (Colgan, 1919).

The coast of Norway from Bergen north to Tromsoe in northern Norway has had a long history associated with the occurrence of sea-beans. The Lofoten Islands opposite Narvik, Norway may be one of the best locations to find sea-beans and other kinds of tropical drift. Sea-beans have been found at the North Cape on the island of Mageray and eastward along the shores of Barents Sea. The southern coast of Norway from Bergen southward appears to be outside the range of most Gulf Stream drift. This also seems to be largely true of continental coastlines from Denmark southward. Guppy (1917), without offering proof, stated that there are tropical drift records for the Jutland coast of western Denmark, and he reported that a sea heart was found in the vicinity of Boulogne, northern France. Leenhouts (1968) reported tropical seeds said to be *Mucuna urens* from the West Frisian Islands in the Netherlands.

The Northeast Atlantic Current is responsible for the tropical debris that reaches the Faeroe, Shetland and Orkney Islands, the Outer Hebrides, the western coasts of Scotland and Ireland, and the beaches of Cornwall and Devon in southwestern England. Western beaches fully exposed to the storms of the North Atlantic are the most productive in this region. This explains the long sea-bean history of this area. If any beaches are particularly rewarding on these rugged coasts, it is the western beaches of the Outer Hebrides. The Isle of Lewis and North and South Uist in the Outer Hebrides are known for the sea-beans and other tropical debris that strand on their beaches. Inner Hebrides Islands, such as Mull and Skye, may also be reasonably rewarding. Although the western coast of Ireland has not been searched to the degree that the Scottish coast has, both coasts would appear to be good collecting areas. Colgan (1919) has supplied numbers and records from Donegal Bay in the north to County Kerry in the south. The north coast of North Ireland has yielded a few records. In England most sea-beans strand on both sides of Cornwall. Fewer sea-beans are found as one goes eastward along the English Channel. The south coast of Devon has yielded a few records and so has the south coast of Wales. Sea-beans have been found as far to the east on the English Channel coast of England as the Isle of Wight and Portsmouth. The North Sea coasts of England and Scotland appear to be devoid of any promise.

Colgan states that the sea heart (*Entada gigas*) is the most common tropical drift disseminule found on European beaches and is also the most widespread. Other common species are the gray nickernut (*Caesalpinia bonduc*), the true sea-bean (*Mucuna sloanei* and *M. urens*), and the sea purse (*Dioclea reflexa*). The latter, which may be confused with the true sea-bean, is collected somewhat less frequently. Less common species include *Merremia discoidesperma, Manicaria saccifera, Crescentia cujete, Sacoglottis amazonica, Caryocar villosum,* and *Calocarpum mammosum* (Fig. 8). These eleven species can safely be included in a European list of tropical drift disseminules whose presence is attributed to the Gulf Stream. Other species should be looked for. They await only the practiced eye of enthusiastic searchers. Some reported disseminules, like the coconut and cashew, probably were discarded from ships.

After the Gulf Stream divides in the vicinity of the Grand Banks, a southeastern arm, the Portugal Current, flows towards the south coast of Europe and then turns southward and finally to the west to become the North Equatorial Current. Persistent drifters, caught in the Portugal Current, may be carried back to the New World tropics. Some could become stranded in the Azores.

Lying some 1,500 km off the coast of Portugal, the rocky Azores might seem an unlikely landfall for sea-beans. However, the islands have a sea-bean history which has been discussed by Charles Darwin and H. B. Guppy. During a sojourn in the Azores, Guppy (1917) collected a number of species recorded from European beaches, and three unrecorded species. These three are *Sapindus saponaria, Astro-*

Figure 8. Examples of tropical disseminules which are carried by the Gulf Stream from the Caribbean region to beaches of northern Europe. A, *Entada gigas* (sea heart); B, *Manicaria saccifera* (sea-coconut); C, *Caesalpinia bonduc* (gray nickernut); D, *Dioclea reflexa* (sea purse); E, *Merremia discoidesperma* (Mary's-bean); F, *Mucuna* spp. (true sea-bean); G, *Sacoglottis amazonica;* H, *Crescentia cujete* (calabash) (X1).

caryum sp., and a fruit thought to belong to the Juglandaceae. Guppy mentioned the north coast of San Miguel and the western end of Pico as places where he did collecting. The drift disseminules, according to Guppy, were familiar to the people of the coast towns and were often picked out of the water by fishermen.

It seems likely that the Portugal Current would bring occasional sea-beans to the Atlantic coast of southern Europe. However, no records have come to our attention from this region. This may reflect a lack of exploration rather than an absence of sea-beans.

Eastern United States

Though seldom mentioned in the literature, eastern United States beaches receive tropical drift disseminules. While the Florida Keys and the southeastern Florida beaches regularly receive sea-beans, beaches from South Carolina to Massachusetts receive them on an irregular basis. The stranding of sea-beans on these beaches can be correlated with sea storms that drive the sea-beans out of the Gulf Stream. The Gulf Stream is too far out to sea to regularly deposit sea-beans.

A pod of the West Indian locust (*Hymenaea courbaril*), containing viable seeds, was collected from a Martha's Vineyard, Massachusetts beach. Sea hearts and true sea-beans have been stranded on beaches at Nantucket and on Cape Cod, including Wings Neck. Some of these seeds have been tested, and they were viable. These and other records, more fully discussed in Gunn and Dennis (1972a) and Dennis and Gunn (1975), are the most northerly for the United States and for North America. Moving southward, the next two records are from Maryland and Virginia. A *Calocarpum mammosum* seed was collected from the beach at Ocean City, Maryland, and an endocarp of the ivory nut palm (*Phytelephas macrocarpa*) was collected from Cape Charles, Virginia. South of Virginia, sea-bean records become more numerous. We have identified 22 species of sea-beans stranded along the Carolina coast. Among the more productive beaches are Cape Fear and Bogue Banks, North Carolina and Myrtle Beach, Huntington Beach, and Pawleys Island, South Carolina (Gunn and Dennis, 1972b).

The Gulf Stream daily deposits large numbers of tropical drift disseminules on the beaches of eastern Florida, especially those from

Cape Canaveral south through the Florida Keys. The beaches of southeastern Florida receive about the same number of species as are found on beaches of Malaysia (Gunn, 1968). Collectors should bear in mind that some beaches may not have this wide a range of sea-beans, because of the activities of man. Beach cleaning operations, coupled with competition from curio seekers, make some beaches less desirable than others. This problem may be overcome if one is willing to go to the trouble of obtaining permission to visit beaches that are less accessible to the public. We have often done this and can report excellent results, especially in the Palm Beach region. Our list for the Palm Beach area alone exceeds one hundred species. The Florida Keys also provide excellent beaches for sea-bean hunting.

☙ Gulf of Mexico

After passing through the Yucatan Channel, the Yucatan Current divides. One arm moves toward the Florida Straits and the other toward the western Gulf. There is sufficient circulation within the Gulf to bring sea-beans, sometimes plentifully, to beaches from Santa Rosa Island in western Florida to beaches at the western edge of the Gulf (Gunn and Dennis, 1973). The western coast of peninsular Florida, on the other hand, has little to offer in the way of sea-beans. The currents are not close enough to the shore.

The best sea-bean collecting on any Gulf Coast beach within the United States is at Padre Island, Texas. This sandy barrier beach, some 190 km in length, is apt to receive rich deposits of tropical debris from late March into early summer. In late March, 1971, we collected some 40 species on a northern beach of Padre Island.

Southward along the Mexican Gulf Coast, we have obtained good results at beaches near the port of Veracruz and near Progreso, Yucatan. Numerous samples have been received from Joann Andrews, who is making a careful collection of the sea-beans along the coast of the Yucatan. Not only do the currents bring sea-beans from distant sources to the Gulf Coast of Mexico, but the tropical vegetation of Mexico is a source. Fruits of *Andira galeottiana*, an endemic Mexican species, were found stranded along Yucatan and Texas beaches (Gunn and Dennis, 1973).

West Indies

With the history of sea-beans in this region going back to the time of Columbus, the West Indies seem almost synonymous with our subject matter. Although the islands supply most of the sea-beans which reach beaches of the North Atlantic, they may also be way stations for others that have arrived by north flowing currents. Some disseminules may come by way of the South Equatorial Current from the west coast of Africa. Others are native to the coast and large river basins of northern South America and Central America. Whatever the source, the beaches of the West Indies are among the most rewarding in the New World for finding sea-beans.

We are impressed by the number of stranded disseminules that we have collected at Palisadoes Beach opposite Kingston, Jamaica. As Morris (1895) stated after visiting Palisadoes in 1884, the beach is exposed to the full force of the waves and receives "large quantities of wreckage, sea-weeds, and drift-fruit." Other Jamaican beaches are also rewarding. One is likely to find a mixture of specimens, some that are fresh and obviously local in origin, and others that are worn and encrusted with marine organisms. The latter have perhaps arrived from far distant sources. The rivers of Jamaica should also be sampled for the drift that they carry. Guppy (1917) commented upon the rich drift of the Black River in southwestern Jamaica. This is still a good river to visit, and a place to see some of the plants that supply drift disseminules.

Guppy spent much time in the West Indies and made notations regarding the drift disseminules he saw on beaches of Grenada, Jamaica, St. Croix, Tobago, Trinidad, and the Turks Islands. The stranded disseminules are quite uniform among the islands. This is to be expected in view of the common current system that dominates the Caribbean Sea.

Eastern South America

Like the West Indies, much of the eastern coast of South America is favorably situated for receiving tropical drift. Sea-beans and other material could easily reach this coast from Africa. The South Equatorial Current from the west coast of Africa divides in the vicinity of

Natal and Recife, Brazil. The main flow of the South Equatorial Current continues toward the West Indies, skirting the northeastern coast of South America. The southern arm, the Brazil Current, moves southward along the coast of Brazil.

Large rivers, such as the Amazon and Orinoco, definitely contribute to the tropical drift which is carried towards the West Indies. Bates (1863) remarked on the large numbers of sea-coconuts (*Manicaria saccifera*) which he saw floating at sea some 700 km beyond the mouth of the Amazon. The fruits appear on beaches throughout the West Indies and a few reach the coasts of Europe. Sea-beans that travel from the Amazon to the shores of western Europe must complete a voyage of some 15,000 km. On the other hand, sea-beans that reach the coast of Brazil from West Africa have a shorter voyage of about 5,000 km.

Africa

If it were not for a medical doctor with an interest in botany, the beaches of Africa would be virtually unknown. Exploring beaches along the African coast, especially those of South Africa, Muir (1937) brought together a great deal of information on sea-beans. We are in the process of studying the drift seeds and fruits in the John Muir collection which is housed at the University of Stellenbosch and administered by Professor P. G. Jordaan.

We know from Muir's travels to the east coast of Africa that excellent collecting opportunities exist at beaches near Lourenco Marques, Mozambique. The east coast of Africa is favored by ocean currents and receives the contributions of large rivers, such as the Zambezi. Judging from a small collection we received from beaches of the Somalia Republic, there are worthwhile opportunities in the equatorial region of coastal East Africa.

The beaches of South Africa are comparable in some respects to those in the Northern Hemisphere that are served by the Gulf Stream. Although well within the temperate zone, beaches on the south coast of South Africa receive a sizable quota of drift disseminules that are characteristic of more northern parts of the Indian Ocean. Their presence is due to a well defined current system that brings tropical debris southward by way of the Agulhas Current and then westward around the Cape of Good Hope. According to Muir, many of the drift disseminules which strand on the beaches of South Africa probably come from Madagascar.

The Agulhas Current brings a number of sea-beans of Indian Ocean origin into the Atlantic. Among them are such long distance drifters as the sea heart, gray nickernut, sea purse, and true sea-bean. These are species that Muir picked up on the coasts of Southwest Africa after they had been brought northward by the Benguela Current. How many other species may round the Cape and never reach a landfall is not known. It seems likely that a number make this journey but eventually sink to the bottom. The desert coast of Southwest Africa has little to offer other than the few far-ranging species that have been mentioned. Even these, according to Muir, are missing as far north as Angola. Not until the mouth of the Congo River is reached do the beaches become more rewarding. Muir found sea-beans at every step on beaches he visited at Pointe Noire on the Congo coast and at Duala in the Cameroons. On beaches from the Cameroons westward, the assortment was not so great.

Ocean currents appear to play only a minor role in bringing sea-beans to the tropic coasts of West Africa. According to Muir most come from rivers and estuaries and therefore are of local origin. The species that do occur are reminiscent of the American tropics and show fewer affinities with the strand flora of the Indo-Pacific region. The fact that West Africa and the New World tropics share a number of species confirms the strong current tie which is shown in the current map.

The northwestern coast of Africa and the Mediterranean coast are unproductive from the standpoint of tropical drift disseminules. The same can be said of the Canary Islands and Madeira which lie off the Atlantic coast in the path of the southward flow of the Canary Current.

Indian Ocean

With its romantic past, the Indian Ocean region is an inviting place to beachcomb. Its beaches are the only ones in the world that have provided records of the fabulous coco-de-mer of the Seychelles. Among the currents that supply beaches through wide reaches of the Indian Ocean are North and South Equatorial Currents, the Equatorial Counter Current, and the Monsoon Current which seasonally sweep southward from the coasts of India.

Beaches on the south coasts of Sumatera (Sumatra), Djawa (Java), Bali and the Lesser Sunda Islands are probably as rewarding as any to be found on the perimeter of the Indian Ocean. Not only are the currents favorable in this region, but Indonesia is generally regarded as the epicenter from which many tropical plants have spread through drifting to other parts of the world. Lists of drift disseminules have been compiled by Guppy (1890) for the Cocos-Keeling Islands to the south of Sumatera and by Schimper (1891) for the south coast of Djawa. David Fairchild (1931) stated that in all his travels he had never seen such masses of drifted tropical seeds as he saw on the beaches of Pulu We, a tiny island lying at the northernmost tip of Sumatera. He attributed this richness to an unusually favorable location in regard to ocean currents.

Krakatau, an island about 50 km from Sumatera and Djawa, lost most, if not all, of its vegetation during a volcanic eruption the last week of August, 1883. The first visitors to Krakatau reported that all of the vegetation had been killed. While this may or may not be accurate, Treub (1888) found 15 species of spermatophytes during his visit in June, 1886. He also noted stranded disseminules of these species: *Barringtonia asiatica, Calophyllum inophyllum, Cerbera odollam, Cocos nucifera, Erythrina* sp., *Heritiera littoralis, Hernandia nymphiifolia, Ipomoea pes-caprae, Scaevola koenigii,* and *Terminalia catappa.* The only disseminule not covered in this book is *S. koenigii.* Docters van Leeuwen (1929) observed that all of the listed species had become an integral part of the island's flora. In summarizing the flora of Krakatau, Ridley (1930) thought that about one half of the flora was introduced by drifting, a quarter by wind, and the remainder by birds and bats.

৺§ Australia and New Zealand

Although some of the drift disseminules of New Guinea and nearby islands have been described by Hemsley (1885), there are only a few scattered references to those found on Australian beaches. The east coast of Australia from the Great Barrier Reef to the vicinity of Sydney is favorably situated for receiving drift which has floated from the hundreds of small islands that lie near the equator in the South Pacific. First carried westward by the South Pacific Equatorial Current, then southward by the East Australia Current, sea-beans are brought close to the Australia coast, and some would be stranded.

The south coast of Australia and the western coast facing the Indian Ocean do not appear to be well situated for receiving tropical drift. The currents are from the south and are unlikely carriers. However, Guppy (1917) reported two gray nickernuts found in 1912 on the shores of South Australia.

The East Australia Current veers eastward toward New Zealand after passing along the east Australia coast. As indicated by occasional reports of tropical drift disseminules from the northernmost beaches of New Zealand, a few more seaworthy drifters make this journey. Among those recorded at Ninety Mile Beach at the northern tip of North Island are the coconut, gray nickernut, sea heart, candlenut, box fruit, and *Mucuna gigantea* (Mason, 1961). Ninety Mile Beach appears to be the only location for finding tropical drift in New Zealand. Other good opportunities exist on the beaches of the Kermadec Islands to the north of New Zealand (Sykes and Godley, 1968). The occasional appearances of tropical drift disseminules in the New Zealand region are related to sea storms.

Eastern Asia

The North Equatorial Current is the principle agent that brings tropical drift from more distant parts of the Pacific to the coasts of eastern Asia. Likely coasts for receiving disseminules are those which face the Pacific in the long arc of island archipelagoes that include the Philippines, Taiwan, Ryukyu Islands, and Japan. The less exposed coasts along the inland seas bordering the mainland are not so likely to receive drift material that is carried by the North Equatorial Current or its northward bound extension, the Kuroshio Current. Most southern parts of the Asiatic coast probably receive an abundant drift from local sources. We have a small collection of tropical drift disseminules from Singapore, and Muir (1937) has supplied a long list of ones collected by Kerr on Kaw Tao Island in the Gulf of Siam. However, little is known concerning the occurrence of tropical drift disseminules along the eastern coasts of Asia and information is badly needed.

Pacific Ocean

The remarkable uniformity of the strand flora of islands throughout much of the Pacific is testimony to the effectiveness of currents and to a lesser extent to birds in bringing viable seeds and fruits to even the most remote of the many small islands. Also helpful in this respect are prolonged gales which sweep large sections of the Pacific. During the winter of 1940–1941, strong westerly winds attaining a velocity at times of 55 knots were assumed to have been responsible for bringing large numbers of drift disseminules to the mid-Pacific island of Canton (van Zwaluwenburg, 1942).

Not to be confused with storms is a phenomenon known as the tsunami which is restricted to the Pacific region. The tsunami is a tidal wave which moves at great speed across the Pacific. It emanates from earthquake or volcano epicenters. These tidal waves carry debris of all kinds rapidly from place to place. Unlike normal waves, the tsunami may deposit debris well beyond the back beach. The effect is to transport drift disseminules from the inhospitable beach to low lying interior districts where soil and moisture conditions would better favor germination and eventual establishment of new plant life. The effect of the tsunami in establishing drift borne species on the northern coast of Oahu, Hawaiian Islands is said to be significant (Otto and Isa Degener, personal communication).

As might be expected, the stranded disseminules of various island groups are similar. This has been noted by others, and we have noted this in comparing drift disseminules from Canton Island in the Phoenix Islands with those from Viti Levu, Fiji Islands (Fig. 9) and collections made by Ray Fosberg, Smithsonian Institution, on beaches of several atolls in the Marshall Islands, Aldabra Island, and Guam. Species that are found commonly on beaches in widely scattered parts of the Pacific include *Aleurites moluccana, Barringtonia asiatica, Caesalpinia bonduc, Calophyllum inophyllum, Cerbera odollam, Entada phaseoloides, Heritiera littoralis, Mucuna gigantea,* and *Pandanus* spp. The discovery of a New World species seed, *Merremia discoidesperma,* on Wotho atoll, Marshall Islands is fully discussed in Gunn (1976). Only islands that are reasonably close to

Figure 9. Tropical drift seeds and fruits collected from a Viti Levu, Fiji Islands, beach by Mrs. Clocker in 1972 (X ⅓).

the American mainland contain plants of New World origin. Otherwise insular floras throughout the Pacific stem mainly from the Indo-Malaysian region and therefore are Old World in origin. Reference was made to the drift disseminules of the Ratak Chain by Kotzebue (1821) and to those of Guam by Safford (1905).

The Hawaiian Islands are too distant from other islands and the main current systems to receive many disseminules other than those that are from plants within the island chain. Carlquist (1970) estimated that 14.3 percent of the original flowering plant immigrants to the Hawaiian Islands are clearly adapted to oceanic drift, while another 8.5 per cent may have arrived by rare or freak flotation events. Nevertheless, the beaches are well worth exploring for disseminules of local origin.

Clipperton Island some 2500 km due west of Costa Rica and about 700 km SSW of the nearest land, between Manzanillo and Acapulco, Mexico, regularly receives drift disseminules. While most of the disseminules arrived from Central America via the Northern Pacific Equatorial Current, some may come from Oceania via the much weaker Equatorial Counter Current. Sachet (1962) recorded these stranded disseminules *Astrocaryum* sp., *Caesalpinia bonduc*, *C. major* (probably *C. bonduc*), *Canavalia rosea*, cocoid palms, *Dioclea megacarpa*, *D. reflexa*, *Entada gigas* (probably *E. phaseoloides*), *Merremia discoidesperma* (misidentified as *M. tuberosa*), *Mucuna mutisiana*, *M. sloanei*, *M. urens*, *Sapindus saponaria*, and *Strongylodon lucidus*.

⌇ Western New World

Currents are of little assistance in bringing tropical disseminules to the long coastline of the New World which faces the Pacific. From Alaska to Tierra del Fuego the Pacific coast is largely under the influence of currents that originate at higher latitudes. The Humboldt Current in the south and the California Current in the north have little potential for bringing tropical drift to the coasts that they dominate. There are, nevertheless, in some areas excellent beaches for finding disseminules that are largely local in origin. Johnston (1949) found a wide variety on the beaches of San Jose Island in the Gulf of Panama. Some were local in origin, while others had been brought by currents from nearby islands or mainland coasts.

Transport Currents and Collecting Beaches

We have a few disseminules sent to us by Corinne Edwards from beaches near Puntarenas, Costa Rica. Among them were the distinctive fruits of *Pelliciera rhizophorae*. This species has a limited distribution along portions of the Pacific coast from Costa Rica to Columbia.

Except for accounts by Guppy (1906) there is little information for the Pacific coasts of South America. Guppy found many species amid the drift on the shores of the Gulf of Guayaquil, Ecuador. But south of Ecuador, particularly along the desert coasts of Chile, there was a virtual absence of current borne disseminules.

Far to the north, there is one record of stranded country almond (*Terminalia catappa*) on the Oregon coast (Gibbons, 1967). Other disseminules should be sought on beaches of the Pacific coast of North America. A lengthy route is involved by way of the Kuroshio, North Pacific, and California Currents. But judging from the numbers of sturdy drifters that reach the northern coasts of Europe, there should be others besides the country almond that make this journey.

4
Collecting and Uses

Those who live outside the tropics may not have to travel long distances to collect sea-beans. In the preceding chapter, we discussed the temperate beaches where sea-bean collecting can be turned into an exciting hobby. The following information is designed to aid the hobbiest find, store, and use sea-beans.

Suggestions for searching

These suggestions apply mainly to temperate beaches where an extra effort is needed to locate sea-beans. In the tropics the beaches are often so littered with drift material that suggestions are not needed.

Once a beach has been selected, start searching in one of two distinct zones: the lower strand (Fig. 10) and the upper strand (Fig. 11). The lower strand is where windrows of seaweed and other debris mark the highest advance of a normal high tide. The upper strand, often at the edge of the back beach, is where exceptionally high tides caused by storms have lodged drift material. Whichever part of the beach you select to walk, remember that you will need a bag or other container to hold the sea-beans, a stick for poking, and perhaps a canteen of fresh water. While walking on the wet compact sand is relatively easy, a walk along the back beach in loose sand may be quite tiring. It is a good idea to stockpile collections along the way out so that they need not be carried for the entire round trip.

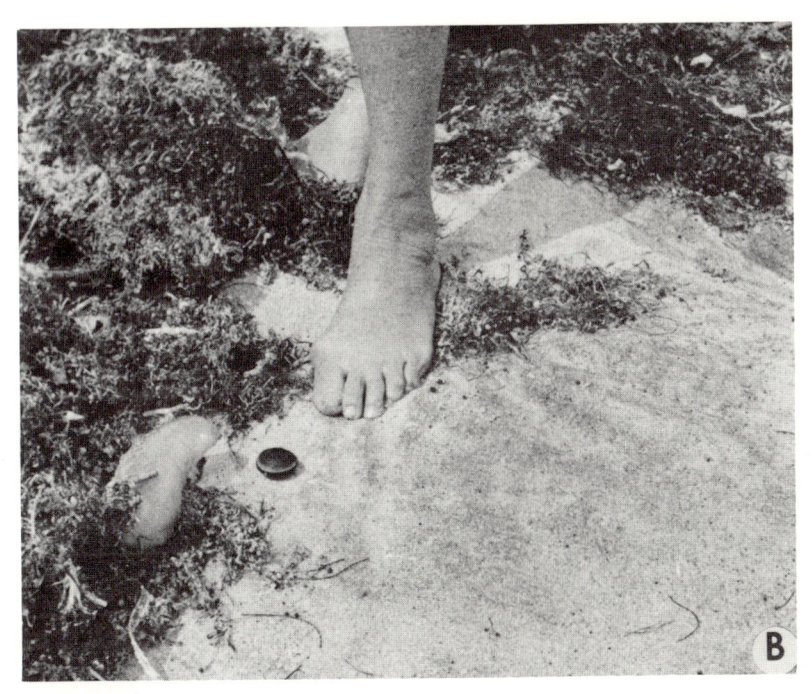

Figure 10. Two views of the lower strand. A, Eleanor Miller and Elvis walking along the lower strand at Riviera Beach, Florida. B, typical lower strand scene (photograph by Robert Mossman).

(43)

Figure 11. Typical upper strand scene (photograph by Robert Mossman).

Collecting and Uses

It is a good plan to begin along the most recent tide line. Unless you are first, you may lose your best trophies to curio seekers and other collectors. Seed collectors, like shell collectors, often find that it is a good policy to make their rounds early in the morning before the beaches become crowded. The best collecting will not be amid the debris of the back beach but where the tides in their twice daily cycle constantly deposit new material. A rising tide near the full stage is the best tide for finding sea-beans. Not only are disseminules likely to strand more frequently at this time, but the lower part of the beach will be covered by water. This will greatly lessen the amount of beach to be searched.

After covering a reasonable stretch of the lower strand, make a return trip by way of the piles of storm debris along the back beach. Huge logs, pieces of lumber, barrels, bottles, and perhaps such reminders of the tropics as coconuts and bamboo may be discovered. It will be difficult to find smaller objects amid this debris. You may want to make many return trips by way of these older deposits that tell of past storms and wind lashed seas.

Do not be disappointed if the sea-beans you find in the litter of the back beach appear less attractive than freshly stranded specimens. The natural lustre can often be restored. Also bear in mind that however broken or battered a specimen may be, it constitutes a record of a drift disseminule far from its native home. A specimen is worth keeping so long as it is complete enough to insure identification.

Do not let seaweed be a nuisance. Usually freshly stranded sea-beans are out in the open and not in the seaweed that sometimes litters a beach. After seaweed has been pushed by the waves to higher levels of the beach, it will lie rotting in long windrows. The time to make a search is after drying and rotting have reduced the bulk so that it is much easier to find embedded objects. The most difficult problem will be with the sand, not the seaweed. A strong wind blowing across a sandy beach can cover over objects almost as quickly as they are stranded. This is good reason to appear promptly on the beach whenever weather and tide conditions seem to be most favorable.

Wind may be an ally by uncovering sea-beans that have previously been buried in the sand. Wind direction and velocity are the most important factors when it comes to the stranding of objects of all kinds. With an offshore wind, watch for incoming debris from the nearby shallow bottom. Shells, sand dollars, sea urchins, and occasional seeds and fruits that have become water-logged and gone to the bottom are objects that may be found under conditions of an offshore wind. Much more productive for the seed collector is an offshore wind of good velocity that holds for at least a day, preferably much longer. On the southeast coast of Florida, where the Gulf Stream comes within three to five km of land, it usually takes a brisk easterly or northeasterly wind to move floating objects shoreward. Better yet is a hurricane that passes some distance out to sea. The same conditions that favor the stranding of sea-beans in Florida apply to the rest of the world.

Identification

Once back from the beach, spread everything on newspapers. The removal of barnacles and other marine organisms (Fig. 12) is a needed precaution if unwelcomed odors are to be avoided. After the specimens are thoroughly dry, sort them. Store similar sea-beans in separate containers. Do not worry about being too accurate in this initial sorting.

Species identification (sometimes we must be satisfied with a generic identification) is essentially a process of elimination. With the help of a key and illustrations, we eliminate the species or genera that do not fit. We either arrive at the correct identification, or we are left with several possibilities. The final stage in making an identification may sometimes require painstaking effort. All the distinguishing features of the disseminule we have before us need to be carefully noted. A hand lens is essential. Also it is helpful to have an assortment of the specimens we are trying to identify. It should be recalled that long immersion in seawater may cause pronounced changes in appearance. Eroded or weathered specimens may look altogether different from fresher ones. Note differences between weathered and slightly weathered specimens in the illustrations of manchineel (*Hippomane mancinella*) and country-almond (*Terminalia catappa*). Our illustrations, except where noted, depict drift disseminules which have undergone varying periods of immersion in seawater.

Do not expect to identify every specimen. Some of our unknowns have been studied by several experts, and the disseminules remain

Figure 12. Tropical disseminules bearing marine organisms (X1).

(47)

unidentified. Disseminules of some genera, like *Erythrina, Ipomoea,* and *Mucuna,* are difficult to identify to species. Growing a plant may be the only way to make an identification. Disseminules which are not true drifters (drift for one month or more) may also be found on beaches. These disseminules may have floated only a short distance or come from the local vegetation.

Do not depend solely on external characters. In some species the internal structure is useful. Sectioning may sometimes be accomplished with a knife blade, but usually a hacksaw with a fine-toothed blade is best. Cross sections of most disseminules are illustrated in the catalog. Cross section characters should be considered in making identifications of drift disseminules.

⇜ *A permanent collection*

We recommend establishing a reference collection in which each specimen is identified and filed in permanent containers. You will need shelf space and separate containers for each species or genus. One of the best methods of housing is in rigid clear plastic containers, about the size of shoe boxes, with tight lids. This kind of container is especially recommended where mice are a problem. Less expensive but adequate containers are clear polyethylene bags. Place bags in an alphabetical order in boxes, drawers, or shelves. It is handy to have a box of unknowns for each collecting locality. These unknowns should be reviewed from time to time as your ability to identify specimens improves. It is always encouraging to note a decrease in number of unknowns.

Fleshy genera, such as *Avicennia, Crinum,* and *Mora,* should be preserved in tightly sealed glass jars containing 70 percent methyl or ethyl alcohol and 30 percent fresh water. Each container, whether sealed jars or boxes, should bear a label giving the common and scientific names (if you know them), the date the material was collected, and the locality. Whenever possible give such details as the name of the beach and distance from the nearest town.

⇜ *Tips for the foreign traveller*

We have already discussed where to collect in Chapter 3. Now to think of steps to be taken before departure. Procure a reliable map,

Collecting and Uses

and make sure that the locality you have chosen is conveniently located so far as beaches are concerned. Consult a travel agent or tour guide about matters concerning lodging, transportation, and the like. Additional information on such subjects can be obtained at an appropriate consulate or embassy. There are naturalists the world over who are intimately acquainted with the seashore and what it has to offer. Check with museums and natural history societies for the names of such persons and write to several.

Never assume that you have exhausted all sources of information. Begin making more inquiries after you arrive at your destination. Be sure to have a copy of this book handy. Turn to the illustrations whenever someone is in doubt about your mission. If you are in a sizable community, chances are good that there will be an institution of higher learning with a botany department and perhaps a museum that employs a botanist. Inasmuch as your hobby comes under botany, you may receive as much help and direction from botanists as from anyone. Do not overlook the amateur shell collector. Few others will have as much information about tides, winds, beach conditions, and the best places for the stranding of drift material. Remember that a good way to return a favor is to keep a supply of sea-bean artifacts with you while travelling.

As you do your collecting, bear in mind that whatever you take back with you must pass through customs. So far as the United States is concerned, you will have little difficulty so long as you bring home specimens for your personal use. There is a limit to how many you can bring in and still claim that you are an amateur collector. Also there are three important rules to observe.

1. All your specimens must be thoroughly dried.
2. No fleshy or pulpy fruits are allowed.
3. Voluntarily show all your specimens to the inspector.

Travellers from other countries should check customs regulations of their governments to see what rules, if any, apply to the entry of tropical drift disseminules.

⇜§ Enhanced beauty

Most sea-beans are collected as natural oddities, and some species are often used as lucky pieces. The following discussions cover two uses which extend beyond merely saving natural sea-beans. One satisfying hobby is sea-bean polishing. Polished sea-beans make excellent jewelry. Another hobby, which is equally fascinating, is the germination of sea-beans and the raising of the ensuing plants.

While sea-beans fresh from the beach have intrinsic beauty, this beauty may be enhanced by cleaning and polishing. It is desirable to retain some sea-beans in their natural state, even to keeping some with encrusting marine organisms (Fig. 12). The process of cleaning sea-beans is simply one of removing oil and dirt by washing them in soapy water, or when necessary in alcohol. Encrustations may be removed by scraping with a knife blade. The degree to which specimens are cleaned depends on their use. We believe that most cleaning should be kept to a minimum. Displays should include sea-beans with their natural surfaces as well as those with cleaned surfaces to better show their natural beauty.

Sea-bean polishing is a more involved process. The first consideration is that most sea-beans cannot be polished. The majority of those that can be polished belong to these genera: *Acrocomia, Aleurites, Astrocaryum, Caesalpinia, Canavalia, Dioclea, Entada, Erythrina, Intsia, Merremia, Mucuna, Oxyrhynchus, Sapindus,* and *Strongylodon* (Fig. 13). Some cocoid palms can also be polished. The purpose of polishing is to remove the scurfy outer seed coat, when present, and then to polish the bony surface. The resulting polished surface exhibits the rich colors and natural lustre of the sea-bean (Fig. 14).

Coe (1894) presents an interesting account of sea-bean polishing in Florida. While Figure 15 is not one of his illustrations, this figure succinctly summarizes the interest in sea-bean polishing in Florida at the turn of this century. The methods which are currently being used by artisans are little different from those which Coe described. The three main methods of polishing sea-beans are: hand polishing with sandpaper, hand polishing with a Cabachon machine, and machine polishing with a rock tumbler.

John R. Bonney obtains excellent results using three grades of Carborundum paper: 300, 400, and 600. The Carborundum paper is cut into 5 cm wide strips, and each strip is folded back 1 cm on itself. The fold at the top of the strip forms a sharp edge that can be used to smooth out irregularities. When the first fold begins to lose its abrasion, another 1 cm fold should be made. Initial or rough polishing is done with the coarse 300 grade. Final polishing is done with the 400 and 600 grade papers. The reverse side of the used strips may be used for final buffing. The work is very slow, about five

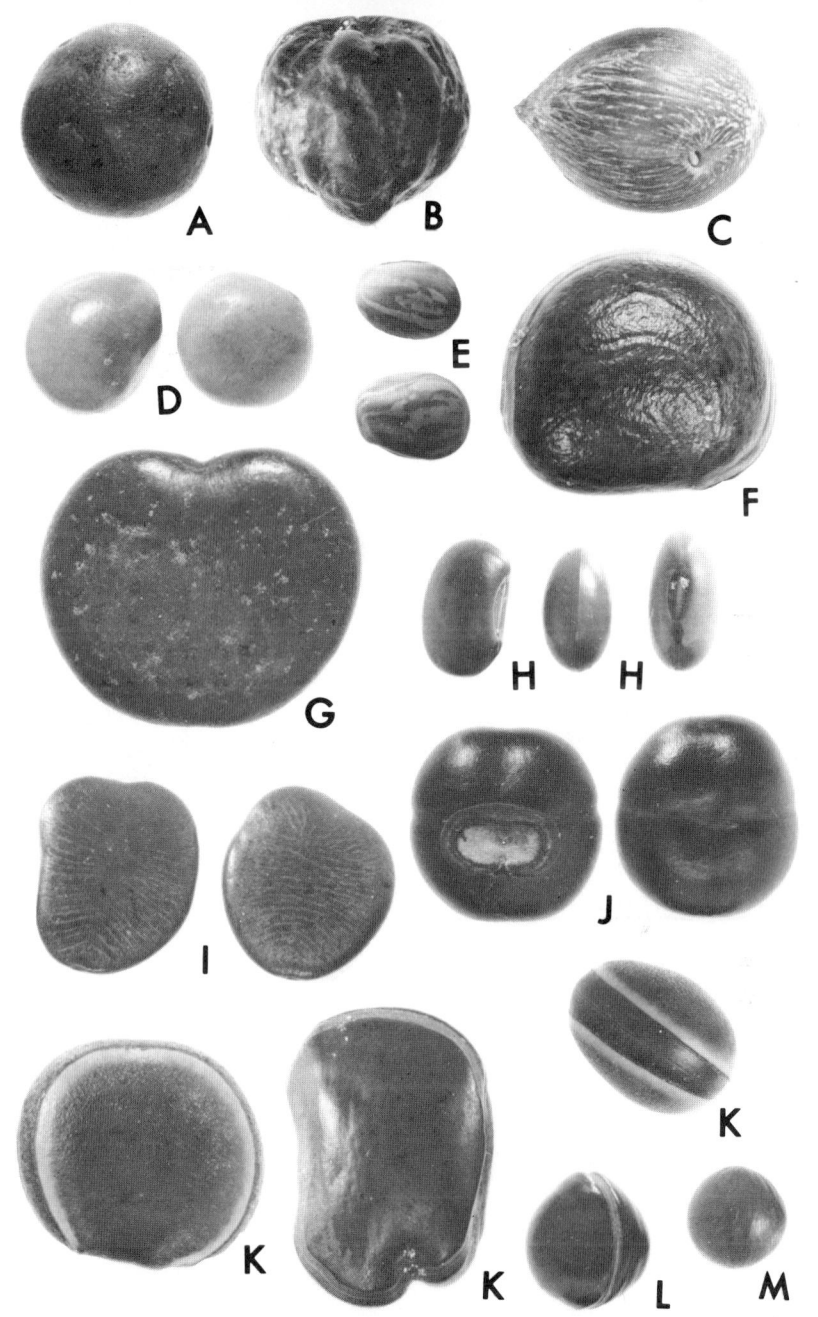

Figure 13. Some of the seeds and endocarps which may be polished. A, *Acrocomia* sp. (prickly palm); B, *Aleurites moluccana* (candlenut); C, *Astrocaryum* sp. (starnut palm); D, *Caesalpinia bonduc* (gray nickernut); E, *Canavalia rosea* (bay-bean); F, *Dioclea reflexa* (sea purse); G, *Entada gigas* (sea heart); H, *Erythrina* sp. (coralbean); I, *Intsia bijuga;* J, *Merremia discoidesperma* (Mary's-bean); K, *Mucuna* spp. (true sea-bean); L, *Oxyrhynchus trinervius;* M, *Sapindus saponaria* (black pearl) (X1).

(51)

Figure 14. Seeds in the left column have their natural surface intact, while those in the right column have been polished. The top and bottom rows are *Mucuna* spp. and the middle row is *Dioclea reflexa* (X1½).

(52)

Collecting and Uses

hours for one sea heart. The reward is a hand polished "gem." For those with time or the desire to keep their hands busy when watching television, try polishing sea-beans by hand.

Much less time is required when the polishing is done on an electric Cabachon machine, equipped with four polishing wheels. C. B. Jessen is shown in Figure 16 using the Cabachon. The outer portion of the seed coat is removed by an emery coated wheel that is cooled by dripping water. Sandpaper on the second and third wheels prepares the surface for polishing. The fourth wheel is a polishing wheel. Its cloth-covered surface is treated with white crocus rouge. During the different stages, the sea-bean is held in place by a dop stick about 20 cm long. The sea-bean is secured to the dop stick by jeweler's wax. After one side has been finished, the sea-bean is removed, reversed, and reattached. Never hold a sea-bean by the fingertips when using this machine. Final buffing should be done by hand. Mr. Jessen has polished several thousand sea-beans and occasionally loses one by cutting through the seed coat.

The tumbler method does not give as good a result as the two methods described above. Its one advantage is that it is automatic, because the seeds are rotated in an electrically operated cylinder. The tumbling time varies with the species of sea-bean. While we cannot give approximate times for tumbling different species, we have found out that different species should not be tumbled together, because of differences in hardness. Tumbled seeds are usually unevenly polished, and some seeds may be cracked or broken. Some hand polishing is required for best results.

৺§ Sea-bean jewelry

While polished sea-beans make fine keepsakes just as they are, most are used in sea-bean jewelry. The various sizes, shapes, and colors lend themselves for uses as earrings, bracelets, necklaces, pendants, and other adornments. For men, many are used on bolo-tie slides and on key rings. The most appropriate use for any sea heart is as a pendant to be worn by a loved one (Fig. 17).

Before a sea-bean can be used for any of these purposes, it has to be pierced so it can be attached singly or with others. One way to make a hole is to use a heated needle that will burn its way through. A more efficient way is to use an electric drill. The seeds may be

Figure 15. A jeweler's advertising clock noting that sea-beans would be polished and mounted owned by Jerry Rich (shown here).

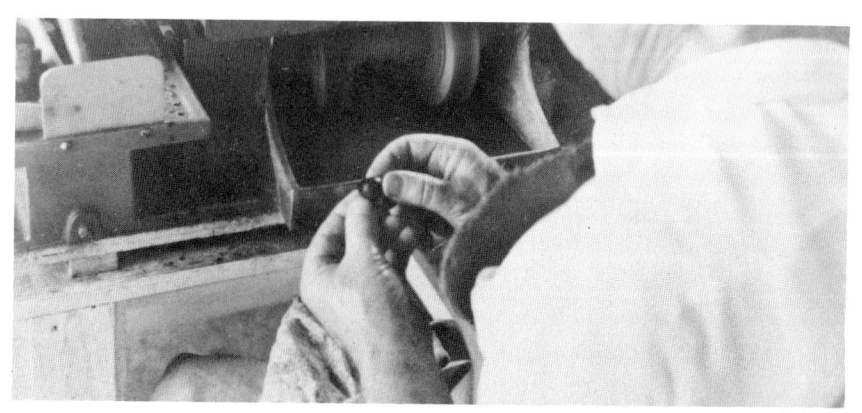

Figure 16. C. B. Jessen at the Cabachon machine. Note the true sea-bean which he is examining.

Figure 17. Sea-beans polished and mounted by C. B. Jessen.

strung, or a screw-eye attached. Sea shells are often intermixed with sea-beans, and polished and unpolished sea-beans may be intermixed.

Unfortunately the effort that has gone into better polishing has not been matched when it comes to clasps or mountings of various kinds. Instead of silver that was used during the last century, modern workers are content with ready-made attachments that can be purchased at any store. This is too bad, for with somewhat greater investment, far more attractive and durable ornaments could be produced.

⤳ *Instructions for growing*

For some, the challenge is how to germinate sea-beans. Not only is it a matter of scientific interest to find out if a sea-bean is still viable after months or years afloat, but interesting plants can often be raised from seeds that have been properly treated. Experiments in germinating sea-beans that have stranded on European shores go back to the time of Linnaeus. Brown (1818) writes: "Sir Joseph Banks informs me that he received some years ago the drawing of a plant, which his correspondent assured him was raised from a seed found on the west coast of Ireland, and that the plant was indisputably *Guilandina bonduc* (*Caesalpinia bonduc*). Linnaeus also seems to have been acquainted with other instances of germination having taken place in seeds "thrown on shore on the coast of Norway." This was a matter that did not escape Charles Darwin's curiosity. On receiving a number of stranded seeds of *Entada gigas* and *Mucuna urens* from the Azores, Darwin sent them to Joseph Hooker at the Royal Botanic Gardens, Kew for planting. The seeds germinated and produced mature plants. Not all records of viable sea-beans on European beaches are old (Clough, 1969 and Gilbert, 1969).

Drift seeds which are likely to be viable are intact and sound in appearance. Most of the viable seeds belong to the legume family, while most palm endocarps do not contain a seed. These two families produce many of the long range drift disseminules. Once a seed has been selected, be sure to scarify the seed coat to permit fresh water to enter the embryo. The solid seed coat which protected the embryo from the lethal seawater now serves as a barrier to fresh water.

Collecting and Uses

Scarification is best done by using a triangular file or a fine toothed hacksaw and wearing a notch into the seed coat until the white embryo is visible. The notch need not be large but it must be deep enough to penetrate the seed coat. Sandpaper may also be used, and a hammer, if there are plenty of seeds to permit a few to be broken. Place the scarified seed in fresh water, and it will imbibe and swell. After soaking for 24 hours, remove and plant in good potting soil at a depth of from 3 to 5 cm. Water the soil, and water as needed thereafter. Seedlings of the sea heart, sea purse, and true sea-beans will emerge in about a month.

Once the seedling has emerged, protect the young plant from temperatures which fall below 40° F. Mature plants may be able to withstand a slight freezing temperature, but these are tropical plants whose spread is limited by freezing temperatures. Many of the legume drift seeds are produced by trees and high climbing vines. Therefore, if the plant is to be a house or greenhouse plant, it should be handled as a bonsai. A viable seed of the West Indian locust tree (*Hymenaea courbaril*) in a drift pod from a beach at Martha's Vineyard, Massachusetts produced a fine tree which made into a bonsai. The plant has remained alive for more than 15 years and has become a popular conversation piece. While all plants grown from viable drift seeds may not be handled this way, this should be considered. The alternative is to have a rapidly growing plant which will cover everything in sight. Stems of the true sea-bean and the sea purse have been measured to grow at a rate of 15 to 30 cm a day in Florida. These plants can only be grown outside in places were freezing temperatures seldom occur.

5
Systematic Descriptions and Illustrations

While we have not been able to identify all our drift seeds and fruits, we have identified the common and unusual ones and assembled them in an illustrated catalog. Disseminules unknown to the reader may be identified by using the following key or by consulting the illustrations and accompanying text.

The catalog illustrations were specially prepared for this book from seeds and fruits selected to show normal size, shape, and texture variation. Usually the text and facing illustration page contain all of the information about one genus. Occasionally three plates are used to illustrate adequately the disseminules. Data in the text are derived from several years of field experience, a review of the literature, and observations submitted by correspondents throughout the world.

On the text pages, the family name is printed in italics above the scientific name, author, and major common name of each important disseminule discussed. Both entries are centered on the page. The disseminules are presented by families so that related genera with similar disseminules will be near each other in the catalog. The remainder of the text consists of these entries which are explained in the following paragraphs: synonym (omitted when no entry), type of disseminule, description, buoyancy factor, buoyancy test, viability test, known transport current, and an assemblage of interesting observations concerning the disseminules and their parent plants.

One or more synonyms are listed only when they have been used in drift literature. No attempt was made to include all botanical

Systematic Description and Illustrations

synonyms. Our nomenclature follows Adams et al (1972) where possible, except for *Canavalia*.

A disseminule may be a fruit, a fruit segment, a mesocarp, an endocarp, or a seed. These terms are pictorially defined in Fig. 78, using a coconut as a model. An entire fruit is encased in an outer thin fruit layer called the exocarp. This rather fragile layer is often eroded in drift fruits, thus exposing either a mesocarp or an endocarp. The mesocarp layer is composed of relatively soft, often buoyant, fruit tissue that can be dented by fingernail pressure. The endocarp layer is composed of bone hard fruit tissues that cannot be dented by fingernail pressure. Not all fruits have both a mesocarp and an endocarp layer. Some have both layers, while others have only one. All fruits originally possessed an exocarp. A few fruits like the two *Entada* species break apart into waterproof segments or compartments. While these segments are not complete fruits, they function as complete fruits. The terms fruit and seed are discussed in Chapter One.

The disseminule description follows this format: scientific name if more than one species involved, type of disseminule, position in illustration, size (length, width, and thickness), outline shape, cross section shape, color, surface topography, and notes as needed about the hilum, fruit scar, or other features.

Buoyancy factors are classified and discussed in Chapter One. Major buoyancy factors are space within a watertight layer and buoyant tissue within a watertight layer. Disseminules are shown in cross section to illustrate the buoyancy principle.

Buoyancy tests were conducted on all disseminules which were available at the time the tests were conducted. The primary source of the tested disseminules was the beach at Palm Beach, Florida. Occasionally fresh disseminules, collected from plants, were used to supplement drift material. The recorded time span is the longest period that at least one disseminule floated. Thus this time span should be regarded as an upper limit and not as an average time. No flotation test exceeded two years. The disseminules were floated in seawater kept in beachcombed one gallon plastic jugs. The jugs were inspected daily, and disseminules which had sunk were removed and appropriate notations recorded. Usually after the internal parts of a disseminule rotted, the remainder floated. Occasionally a sound viable disseminule would sink. Some soft fruits became putrid while still floating and were removed before their duration was determined. Each week the seawater was replaced, and the disseminules washed in a strong jet of fresh water to remove the accumulated slime. The fresh water was drained prior to replacing the disseminules in new seawater.

Most of the viability tests were made using 2, 3, 5-triphenyl tetrazolium chloride, commonly called tetrazolium. Seeds were scarified, soaked in fresh water for at least several hours, and then placed in a one percent aqueous solution of tetrazolium. The colorless tetrazolium solution is oxidized in the presence of dehydrogenase enzymes to its colored (red) elementary form. Because live seeds must have active dehydrogenase enzymes, their embryos turn a bright carmine red. Embryos of dead seeds remain their natural color. The tetrazolium tests were supplemented by some germination tests and beach observations of germinating disseminules.

The known carrier currents are listed. This should not be considered a definitive list. It is only offered as a guide.

The general discussion is an assemblage of facts about the plants which produce the drift disseminules. Emphasis is placed on the role drifting has had in the spread of the species, uses of and oddities about the parent plant and disseminule, and interesting notes about the disseminule.

The illustrations were specially prepared for this book and were made from selected drift specimens, except where noted. The specimens were selected to show range of size, shape, and texture. The cross section is presented to illustrate the buoyance principle.

Disseminule key

To use the following dichotomous key, consider each contrasting pair of statements which bear the same number. Only one statement can be correct. Decide which one is correct and continue until an answer in the form of a genus name is reached. The number following the genus name is the page number where the genus is discussed. The description and illustrations should be checked before assuming the answer is correct. Disseminules which do not key out or match the description or illustrations may be ones which are not included in the book. These may be sent to the authors for identification. In

Systematic Description and Illustrations

addition to an unknown disseminule, one should have a millimeter ruler and a sharp knife blade or small hack saw available.

1. Disseminule 10 cm or more long.
    2. Margin bearing a conspicuous notch.
        3. Disseminule more than 20 cm long .... *Lodoicea*, 186
        3. Disseminule less than 20 cm long.
            4. Disseminule earlike, strongly compressed in cross section .... *Enterolobium*, 146
            4. Disseminule not earlike, rounded in cross section.
                5. Notch at right angles to length ...... *Mora oleifera*, 156
                5. Notch at base and may be obscured by tuft of coarse black hairs .... *Borassus*, 176
    2. Margin entire.
        6. Disseminule at least 3 times (often greater) longer than wide.
            7. Disseminule divided into numerous compartments, or strap-shaped and often drifting as single valve.
                8. Disseminule round in cross section .... *Cassia*, 132
                8. Disseminule strap-shaped, often drifting as single valve .. .. *Delonix*, 138
            7. Disseminule not compartmentalized and strap-shaped.
                9. Disseminule a seedling, bearing no seeds .... *Rhizophora* and related genera, 194
                9. Disseminule a pod bearing 1 to several seeds.
                    10. Pod bearing 2 parallel ridges along upper suture .... *Canavalia rosea*, 130
                    10. Pod without ridges.
                        11. Pod indehiscent; from Atlantic Ocean .... *Hymenaea*, 150
                        11. Pod dehiscent; from Pacific Ocean .... *Castanospermum*, 134
        6. Disseminule less than 3 times longer than wide.
            12. Disseminule flattened in cross section.
                13. Disseminule with stout beak .... *Pelliciera*, 206
                13. Disseminule beakless.
                    14. Surface deeply fluted and black .... *Nypa*, 192
                    14. Surface shallowly grooved and tan to grayish brown .... *Mangifera*, 70
            12. Disseminule round in cross section.
                15. Surface prominently ribbed or tuberculate.
                    16. Surface ribbed .... *Grias*, 124
                    16. Surface tuberculate .... *Annona squamosa*, 74

15. Surface smooth.
    17. Disseminule center hollow or nearly so.
        18. Outer coat wall thin, 2 mm or less in thickness.
            19. Disseminule length or diameter 5 cm or more ..
            .... {*Crescentia*, 82
                                          {*Dendrosicus*
            19. Disseminule diameter 4 cm or less ....
            .... *Calocarpum*, 201
        18. Outer coat wall thicker, over 5 mm thick.
            20. Disseminule oblong and somewhat flattened ....
            .... *Hymenaea*, 150
            20. Disseminule ellipsoidal, globose, or top-shaped.
                21. Disseminule top-shaped with square apex ..
                .... *Barringtonia*, 80
                21. Disseminule globose or ellipsoidal with rounded apex .... *Cocos*, 182
    17. Disseminule center filled with winged seeds or dried pulp.
        22. Center filled with winged seeds .... *Swietenia*, 170
        22. Center filled with dried pulp ... *Annona glabra*, 74
1. Disseminule less than 10 cm long.
    23. Hilum linear occupying 75 percent of seed circumference, linear and thick (liplike), or broad and occupying about 20 to 30 percent of seed circumference.
        24. Hilum broad .... *Calocarpum*, 201
        24. Hilum linear.
            25. Hilum thick and liplike, about as long as seed or shorter
            .... *Pangium*, 108
            25. Hilum a colored band, not thickened, occupying about 75 percent of seed circumference.
                26. Hilum 2 mm or more wide .... *Mucuna*, 158
                26. Hilum 1 mm or less wide.
                    27. Seed compressed .... *Dioclea*, 140
                    27. Seed globose.
                        28. Hilum with raised margin; Atlantic Ocean
                        .... *Oxyrhynchus*, 162
                        28. Hilum with flush margin; Pacific Ocean ..
                        .... *Strongylodon*, 162
    23. Hilum or scar not as above.
        29. Disseminule strongly compressed in cross section.
            30. Outline resembling a human ear.. *Enterolobium*, 146
            30. Outline not earlike.

(62)

Systematic Description and Illustrations

31. Surface smooth or nearly so.
    32. Color lustrous chocolate brown.
        33. Surface bearing numerous faint concentric fracture lines .... *Intsia,* 152
        33. Surface without these fracture lines .... *Entada,* 142, 144
    32. Color dull dark brown to ochre or sepia gray.
        34. Length less than 2.5 cm, or a seedling with visible root .... *Avicennia,* 78
        34. Length or diameter 5 cm or more, not a seedling .... *Fevillea,* 96
31. Surface spiny, veined, or grooved.
    35. Surface spiny .... *Caryocar microcarpum,* 86
    35. Surface veined or grooved.
        36. Surface covered by irregular network of veins.
            37. Disseminule irregular rounded in outline .... *Pterocarpus,* 166
            37. Disseminule ellipsoidal or rounded in outline.
                38. Surface bony, not dented by fingernail pressure .... *Carya aquatica,* 120
                38. Surface fibrous, easily dented by fingernail pressure .... *Dalbergia,* 136
        36. Surface covered by parallel closely or widely spaced grooves.
            39. Disseminule with stout beak .. *Pelliciera,* 206
            39. Disseminule beakless.
                40. Disseminule 4 mm or less thick and bearing herringbone pattern ... *Peltophorum,* 164
                40. Disseminule 5 mm or more thick and surface grooved.
                      41. Basal scar large and circular .... *Mangifera,* 70
                    41. Basal scar absent or inconspicuous .... *Terminalia,* 88
29. Disseminule round or nearly so in cross section.
    42. Disseminule bearing 1 to 3 pores or pits.
        43. Pores along equator of endocarp .... *Acrocomia,* 172
        43. Pores basal.
            44. Surface around pores bearing radiate striations .... *Astrocaryum,* 174
            44. Surface not radiate striated .... *Cocos,* 180 and Cocoid palms, 178
    42. Disseminule without pores.

45. Disseminules bearing conspicuous scar or stem remnant.
  46. Disseminule bearing a cross on side opposite scar .... *Merremia discoidesperma,* 94
  46. Disseminule without cross.
    47. Disseminule less than 3 cm long.
      48. Scar basal, large, circular .... *Quercus,* 106
      48. Scar subbasal.
        49. Seed bearing a ridge on the hilum face; some species hairy .. *Ipomoea,* 115 and *Merremia tuberosa,* 94
        49. Seed smooth on hilum face .... *Mastichodendron,* 202
    47. Disseminule 3 cm or more long.
      50. Scar encircled by a thickened ring ... *Crescentia,* 82
      50. Scar flush.
        51. Scar bearing stem remnant.
          52. Disseminule top-shaped with square apex ... . *Barringtonia,* 80
          52. Disseminule ovate, pyriform, or elongate with round apex.
            53. Disseminule hollow within .... *Hymenaea,* 150
            53. Disseminule containing many winged seeds .... *Swietenia,* 170
        51. Scar not bearing stem remnant.
          54. Disseminule globose .... *Manicaria,* 190
          54. Disseminule not globose.
            55. Seed lustrous, often brightly colored; hilum with more or less parallel margins and often black .... *Canavalia,* 130 and *Erythrina,* 148
            55. Seed dull brown; hilum circular to twisted ... *Carapa, Phytelephas, Xylocarpus,* 168
45. Disseminule bearing inconspicuous scar, not bearing a stem remnant.
  56. Disseminule either bearing basal involucre, pencillike, or green.
    57. Disseminule green .... *Crinum,* 68
    57. Disseminule not green.
      58. Basal involucre composed of broad bracts .... immature *Cocos,* 182
      58. Basal involucre a cylinder, or if absent, disseminule pencillike ...... *Rhizophora* and related genera, 194
  56. Disseminule without basal involucre and neither pencillike nor green.

Systematic Description and Illustrations

59. Surface smooth or at least most parts of surface smooth.
  60. Surface winged.
    61. Seed less than 2 cm long .... *Annona glabra*, 74
    61. Fruit over 4 cm long .... *Heritiera*, 204
  60. Surface wingless.
    62. Disseminule triangular in cross section .... *Canarium*, 84
    62. Disseminule round or nearly so in cross section.
      63. Disseminule black or blackish brown.
        64. Disseminule more than 5 cm long .... *Cerbera manghas*, 76
        64. Disseminule less than 3 cm long.
          65. Outer coat thick and hard, center open .... *Sapindus*, 198
          65. Outer coat thin, center filled .... *Hernandia*, 114
      63. Disseminule brown or lighter color.
        66. Surface bearing numerous faint concentric fracture lines .... *Caesalpinia*, 128
        66. Surface without fracture lines.
          67. Outer wall about 1 mm thick.
            68. Disseminule filled with pulp and seeds ... . *Annona glabra*, 74
            68. Disseminule a seed .... *Cycas*, 98
          67. Outer wall thicker.
            69. Disseminule globose .... *Calophyllum*, 110
            69. Disseminule ellipsoidal .... *Carya*, 120
59. Surface roughened.
  70. Surface tuberculate, or bearing conspicuous knobs, coarse fibers, or stellate shaped.
    71. Surface tuberculate .... *Annona squamosa*, 74
    71. Surface knobby, coarse fibrous, or stellate.
      72. Surface bearing black knobs .... *Caryocar glabra*, 86
      72. Surface coarse fibrous or stellate.
        73. Disseminule stellate .... *Hippomane*, 102
        73. Disseminule coarse fibrous.
          74. Disseminule more than 5 cm long .... *Cerbera odollam*, 76
          74. Disseminule less than 4 cm long .... *Spondias*, 72
  70. Surface not as above.
    75. Disseminule bearing empty cavities .... *Sacoglottis*, 116
    75. Disseminule without cavities.

76. Disseminule in cross section divided into 2 to 3 compartments, or inner fruit wall lobed.
   77. Disseminule 6 to 7 cm long .... *Canarium decumanum*, 84
   77. Disseminule 5 cm or less long.
      78. Diameter 2.5 cm or more; outer wall rough .... *Juglans*, 122
      78. Diameter up to 2 cm; outer wall relatively smooth .... *Carya*, 120
76. Disseminule entire within and inner wall of fruit smooth.
   79. Disseminules 3 cm or less long.
      80. Disseminule with prominent basal scar .... *Sapindus*, 198
      80. Disseminule without prominent basal scar.
         81. Seed coat wrinkled or rough .. *Aleurites*, 100
         81. Seed coat smooth, adherent fruit tissue gives rough appearance .... *Hernandia*, 114
   79. Disseminule 4 cm long or longer.
      82. Surface fibrous.
         83. Outer wall about 1 mm thick .. *Mammea*, 112
         83. Outer wall over 2 mm thick ...... *Andira*, 126
      82. Surface not fibrous.
         84. Disseminule bearing marginal notch .... *Mora*, 154, 156
         84. Disseminule without marginal notch.
            85. Surface tuberculate ...... *Omphalea*, 104
            85. Surface with well developed walls .... *Calatola*, 118

CATALOG

Amaryllidaceae

CRINUM AMERICANUM L., SOUTHERN SWAMP-LILY
CRINUM ASIATICUM L., ASIAN SWAMP-LILY

Disseminule: Seed.
Description: *C. americanum* seed (A-F) 2.5 to 4 cm long, 1.5 to 2.5 cm in diameter, ellipsoidal to ovate, round in cross section, green, surface smooth between ridges. *C. asiaticum* (G-H) seed 2 to 5 cm long, 2 to 4 cm wide, ellipsoidal to irregularly shaped, compressed in cross section, grayish white, surface rough.
Buoyancy factor: Buoyant seed tissues.
Buoyancy: Not tested.
Viability: Most seeds sprouted while in storage.
Currents: Tropical currents arising in the New World and for *C. asiaticum* southeast Asian and Pacific currents.

Swamp-lily seeds appear to be anomalies among drift disseminules, because they appear to be too fragile to survive in ocean water. However, they are found stranded alive on beaches. While little is known about their drifting capacity, we expect that localized spreading by sea currents and rivers does occur. It is unlikely that these seeds are long range drifters, either dead or alive. The seeds which we have collected and which others have reported to us are probably of local origin. *Crinum americanum*, a native of southeastern United States, is commonly found along muddy banks of rivers and lakes. *Crinum asiaticum* has become well established throughout the tropics and subtropics wherever it has been introduced as an ornamental. Both species are excellent ornamentals and may be easily hybridized. Stranded seeds cannot be stored, except in alcohol, because they germinate in storage. As the seedlings develop, the seeds shrivel until the seedlings die from lack of food. Some of our seedlings have lived for more than one and one half years in storage, because they have been stored in a lighted area.

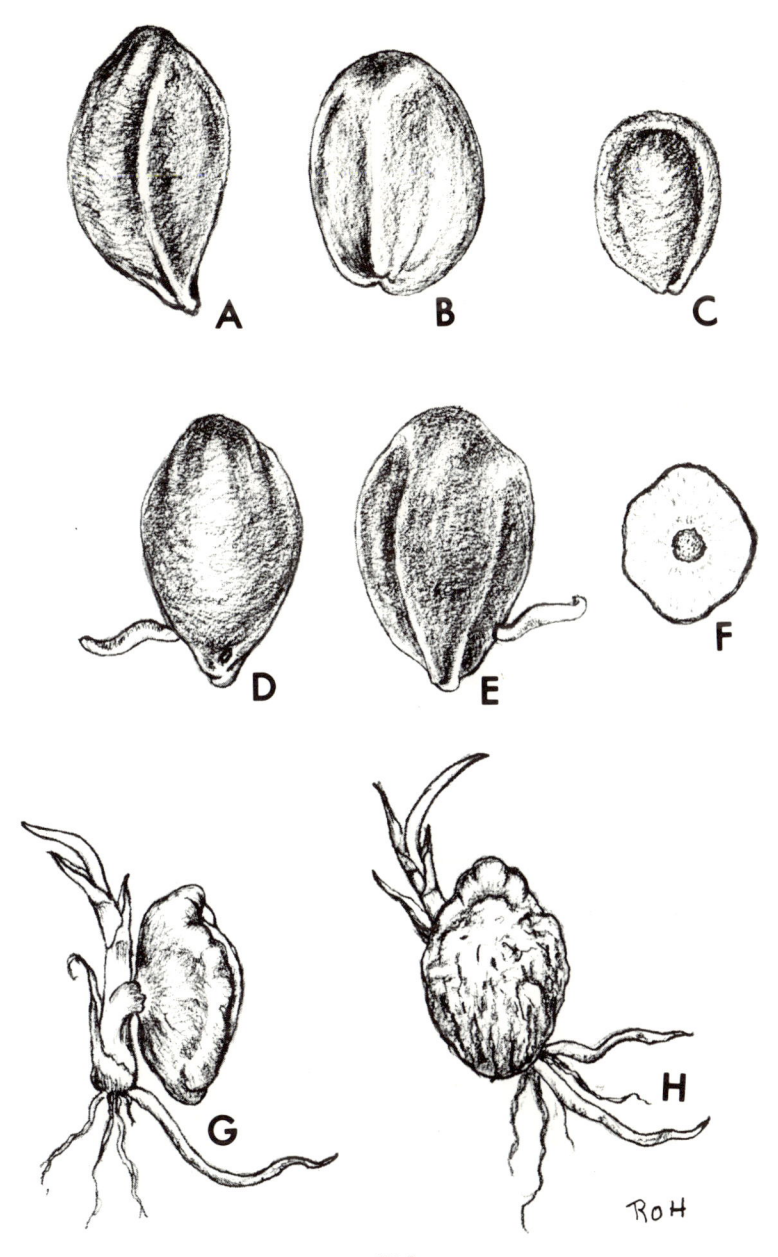

Amaryllidaceae

CRINUM SPP.

Figure 18. *Crinum americanum* (A-F); *C asiaticum* (G-H). A-C, F, seeds; D-E, G-H, seedlings. A-E, G-H, lateral views; F, cross section (X1). *C. americanum*, Dennis, southeastern coast of Florida; *C. asiaticum*, Walls, Grand Terre Island, Louisiana.

Anacardiaceae

MANGIFERA INDICA L., MANGO

Disseminule: Endocarp, rarely fruit.
Description: Endocarp (A-E) 4.5 to 10 or more cm long, 2 to 5 or more cm wide, ellipsoidal, strongly compressed in cross section, tan to grayish brown, surface covered by fibers and bearing well defined parallel grooves from base to apex, basal scar well developed, ellipsoidal and indented.
Buoyancy factor: Endocarp empty or nearly so.
Buoyancy: About 3 months.
Viability: None were found to be viable.
Currents: Any tropical current.

The mango is an important tropical fruit and shade tree that has attained a pantropic distribution through man's activities. A native of southeast Asia, this species was introduced into the Caribbean region about the middle of the eighteenth century (Little and Wadsworth, 1964). Beaches which have received mango endocarps are widely separated. Most authors agree (Guppy, 1917; Muir, 1937; and Ridley, 1930) that stranded records like those from northern European and Australian beaches represent beach garbage derived from passing ships, rather than true drifting. Even Muir thought that the Riversdale, South Africa endocarps could have been beach garbage. Endocarps of freshly eaten fruits may be buoyant. While true for these regions, the endocarps drift and are found stranded on United States beaches as far north as Cape Hatteras, North Carolina (Gunn and Dennis, 1972b). Occasionally entire fruits may be found stranded. They are usually decomposed, resembling a slightly rotten white potato. Such a fruit, which bore a viable seed, was collected on a Florida Key. Because mangos are planted on the Keys, this fruit may have been locally produced.

Anacardiaceae

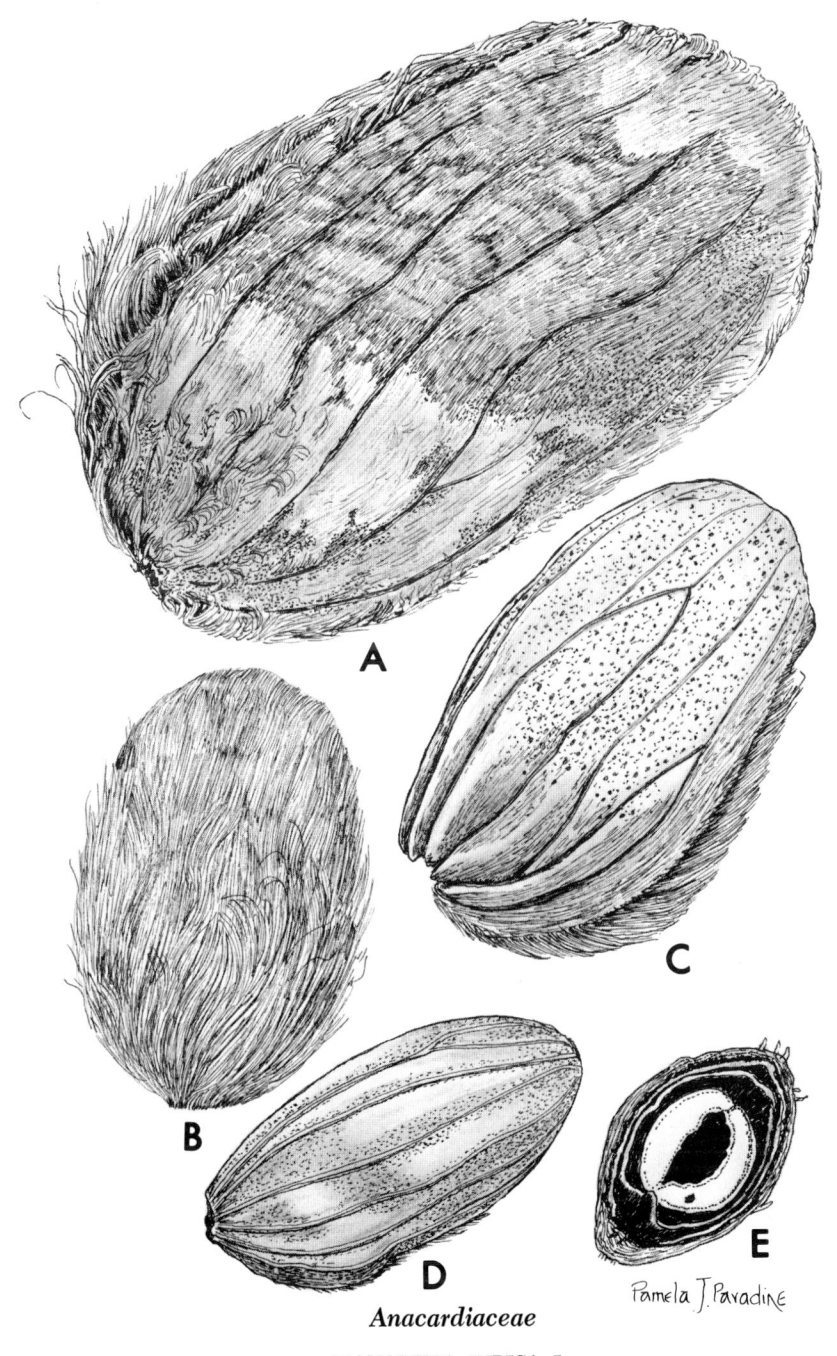

MANGIFERA INDICA L.

Figure 19. *Mangifera indica* endocarps. A-D, lateral views; E, cross section (X1). Endocarps, Gunn and Dennis, southeastern coast of Florida.

Anacardiaceae

SPONDIAS DULCIS S. PARKINSON
SPONDIAS MOMBIN L., HOG-PLUM

Synonym: *S. mombin* is *S. lutea* L. in drift literature.
Disseminule: Mesocarp, rarely fruit.
Description: *S. dulcis* mesocarp (A-C) 2.5 to 3 cm in diameter, globose, round in cross section, tan, surface bearing numerous prominent dark brown fibers which protrude from the mesocarp. *S. mombin* mesocarp (E-I) 1.5 to 3 cm long, 1.5 to 2.5 cm in diameter, oblong, round or nearly so in cross section, tan, surface covered by numerous fine tan fibers with major darker fibers visible on deeply eroded mesocarps (G). Fruit (D) black, thin exocarp seldom present.
Buoyancy factor: Corky mesocarp.
Buoyancy: *S. dulcis*—not tested; *S. mombin*—at least 2 years.
Viability: *S. dulcis*—not tested; *S. mombin*—about 20 percent viable.
Currents: *S. dulcis*—tropical currents arising in Pacific region; *S. mombin*—tropical currents arising in the New World and perhaps the Old World.

Little is known about the drift capacity of the mesocarps of *S. dulcis*. This species, a native of the South Pacific, has been planted in the New World tropics. Thus, mesocarps should be expected on New World beaches.

Spondias mombin mesocarps are common in tropical currents of the New World. However, this species may not be a native of tropical America. Except locally, drifting has played little role in the distribution of the hog-plum. Stranded mesocarps exhibit various degrees of erosion, and deeply eroded specimens (H) never contain viable seeds. Ripe fruits are edible and mainly fed to livestock, hence the common name. It is possible that some of the smaller mesocarps are from a closely related species, *S. purpurea* L.

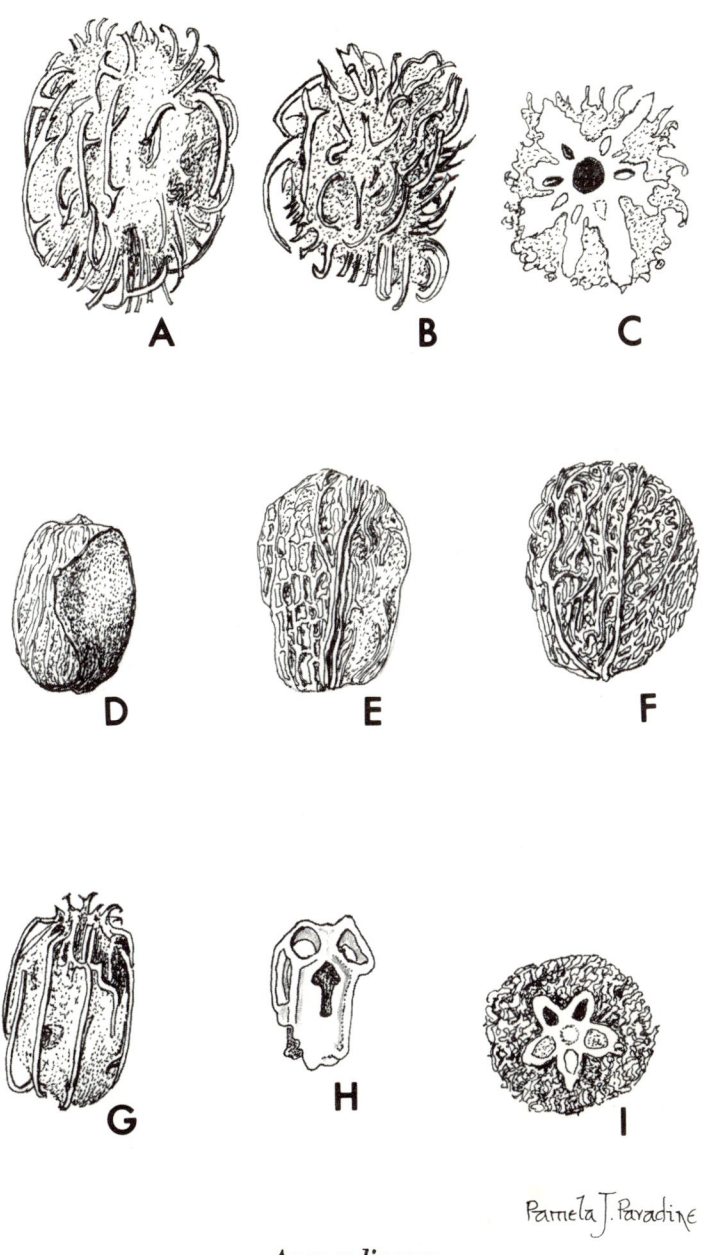

Anacardiaceae

SPONDIAS SPP.

Figure 20. *Spondias dulcis* (A-C); *S. mombin* (D-J). A-C, E-J, mesocarps; D, partially eroded fruit. A-B, D-H, lateral views; C, I, cross sections (X1). A-C, Degener and Degener, beaches of Canton Island; D-I, Gunn and Dennis, southeastern coast of Florida.

Annonaceae

ANNONA GLABRA L., POND-APPLE
ANNONA SQUAMOSA L., SUGAR-APPLE

Synonym: *A. glabra* is *A. palustris* L. in drift literature.
Disseminule: Fruit or seed.
Description: *A. glabra* fruit (A) usually damaged, originally up to 12 cm long and 8 cm in diameter, ovoid, round in cross section, chocolate brown, surface smooth though becoming wrinkled. Seed (B) 12 to 15 mm long, 8 to 10 mm wide, oblong, round at apex and truncate at base, winged, compressed in cross section, dull to lustrous tan, grayish brown, or dark brown, surface smooth. *A. squamosa* fruit (D-E) usually damaged, originally up to 10 cm in diameter, globose to cordate, brownish black, surface composed of a series of overlapping tubercles. Seed (F) about 13 mm long, oblong, compressed in cross section, lustrous dark brown to blackish, surface smooth.
Buoyancy factor: Buoyant fruit tissue; corky seed coat for *A. glabra* seeds; *A. squamosa* seeds are not buoyant.
Buoyancy: *A. glabra* seeds—about 10 months; fruits—not tested. *A. squamosa*—seeds not buoyant; fruits about 7 months.
Viability: *A. glabra*—less than 10 percent; *A. squamosa*—not tested.
Currents: Most tropical currents.

Annona glabra has several common names including pond-apple, alligator-apple, and monkey-apple. The barely edible fruit is produced by trees native to the swamp or marsh regions of the New World tropics. Man has introduced the pond-apple to the Old World tropics. Few fruits strand, because they quickly decay in rivers. Freed seeds which do not germinate in river water may reach the ocean and drift.

Annona squamosa, sugar-apple or sweet sop, fruits resemble unopen pine cones. C. E. Smith, Jr. suggests that the drift fruits are those which dry on the tree (a regular occurrence) and not those derived from fresh, juicy fruits. This explains why so fragile appearing fruit can drift. While the stranded fruits are neither eye-appealing nor edible, tree ripened fruits are a tropical delicacy, eaten raw or used in preparing drinks and sherbets. Like other important tropical American food plants, sugar-apple reached India during the sixteenth century via the Philippines or Cape of Good Hope trade route.

The large *Annona* sp. seed (C) is about 25 mm long, 10 mm wide, is lustrous brown, and has a smooth surface.

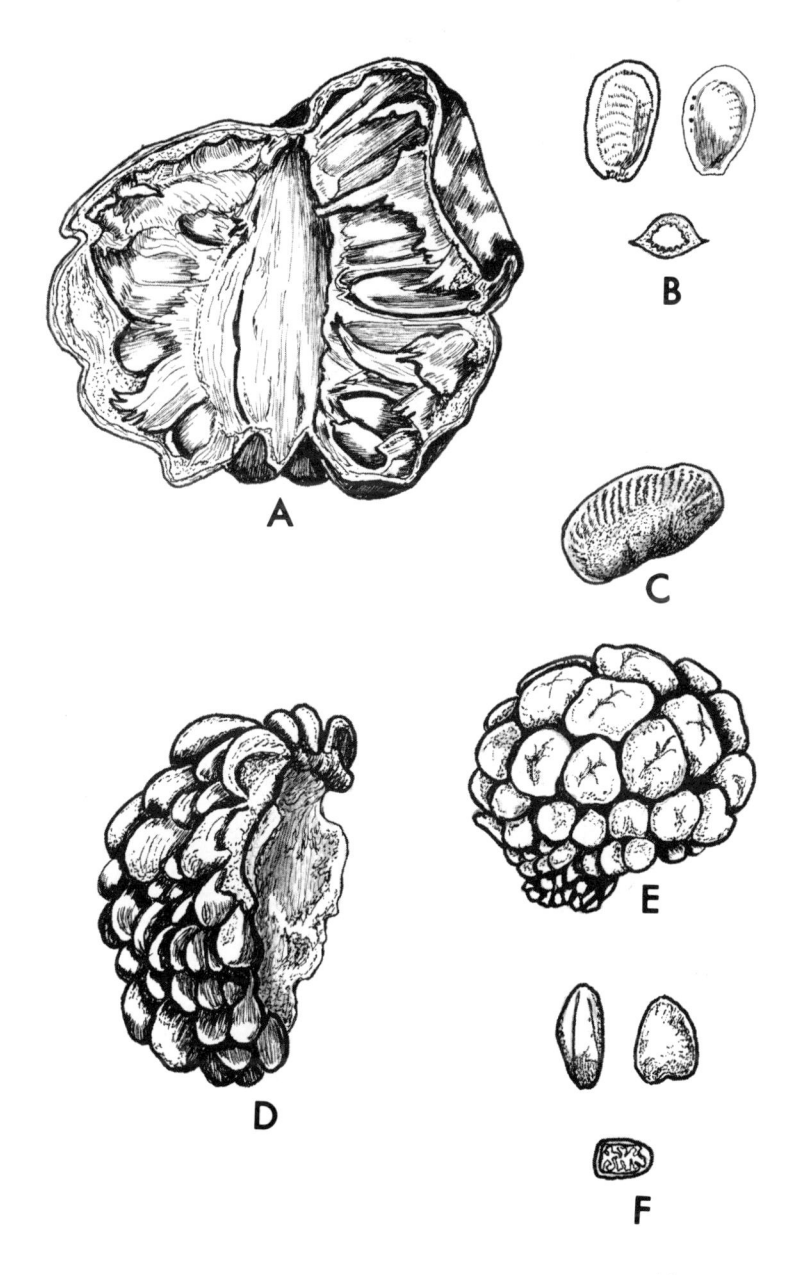

Annonaceae

ANNONA SPP.

Figure 21. *Annona glabra* (A-B); *A.* sp. (C); *A. squamosa* (D-F). A, D, E, fruits; B, C, F, seeds. A, lateral view cut in half; B, F, seeds in lateral and cross-sectional view; C, lateral view; D, damaged fruit; E, fruit (X1). Fruits and seeds, Gunn and Dennis, southeastern coast of Florida.

Apocynaceae

CERBERA MANGHAS L.

CERBERA ODOLLAM GAERTNER

Disseminule: *C. manghas*—fruit; *C. odollam*—mesocarp.

Description: *C. manghas* fruit (A-B) 5 to 7 cm long, 4 to 5 cm wide, ellipsoidal, slightly compressed in cross section, blackish brown, surface smooth. Mesocarp with an outer fibrous layer and an inner corky layer, both layers about 1 cm thick. *C. odollam* mesocarp (C-E) same size and shape as *C. manghas* fruit, but tan and surface composed of well defined fibers surrounding a corky layer, both layers 1 to 2 cm thick.

Buoyancy factor: Fibrous-corky fruit.
Buoyancy: Not tested.
Viability: Not tested.
Currents: Tropical currents arising in the southeast Asia and Pacific regions.

These two species of small trees are native to southeast Asia. While these trees closely resemble each other, they may be identified by the color in the center of the flower. The center is red (orange-pink turning reddish-pink) for *C. manghas* and yellow for *C. odollam*. The natural habitat for *C. manghas* is a sandy or rocky sea coast, while for *C. odollam*, it is a muddy tidal river basin. Fruits have no problem in entering the ocean from either habitat. The fruits are quite buoyant and may drift for long distances. There is no information about how long the seeds remain viable while drifting. Because neither species has attained a pantropic distribution, we assume that seeds in the drift fruits are not long lived. Drifting must have been a factor in the local spread of these species. One reason that the seeds may not live during long term drifting is the thin mesocarp wall that separates the fruit halves. This very thin wall, the straight wall in D, is easily penetrated by seawater, even though the outer mesocarp wall is quite thick.

Apocynaceae

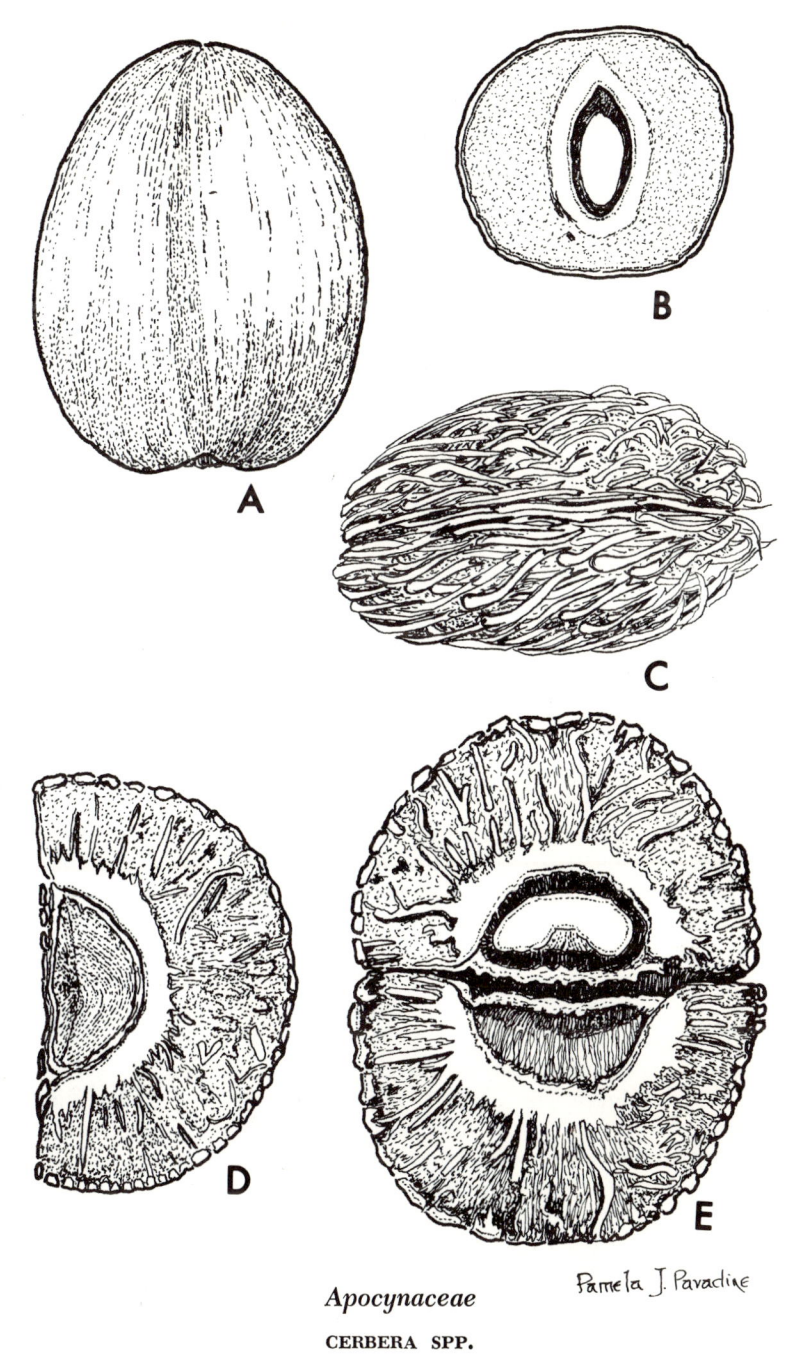

CERBERA SPP.

Figure 22. *Cerbera manghas* (A-B); *C. odollam* (C-E). A-B, fruits; C-E, mesocarps. A, C, lateral views; B, D-E, cross sections (X1). Fruits and mesocarps, Degener and Degener, beaches of Canton Island.

Avicenniaceae

AVICENNIA GERMINANS (L.) L., BLACK MANGROVE

Synonym: *A. nitida* Jacq. of most literature.
Disseminule: Seedling, sprouted fruit, or rarely unsprouted fruit.
Description: Seedling (E-K) up to 12 cm long with a conspicuous hairy root and 2 greenish cotyledons. Sprouted fruit (B-D) up to 5 cm long, 2.5 cm wide, oblong to elliptical, compressed in cross section, dark brown, smooth (hairy when fresh), bearing a protruding hairy root. Unsprouted fruit (A) similar but lacking the protruding root.
Buoyancy factor: Unfolded cotyledons are said to serve as miniature boats, but this does not explain how germinating and non-germinating fruits float. It would appear that buoyancy is due to buoyant seedling or fruit tissues.
Buoyancy: About 1 year.
Viability: Nearly 100 percent of the seedlings were viable.
Currents: Tropical currents arising in the New World and along the west coast of Africa.

Unlike most disseminules, the black mangrove usually drifts as seedlings, not as seeds or fruits. The fruit acts as a surrogate seed coat, because the seed coat is absent. The embryo germinates while the fruit is still attached to the parent tree. When the seedling drops, it may be self-planted in the mud below the parent tree, or be carried into the ocean by the tide. Black mangrove trees are frequent to common members of the tidal swamps along the tropic and subtropic coasts of the New World and west Africa. Black mangrove was introduced into west Africa by man. While the red mangrove (*Rhizophora mangle*) and its relatives have rugged appearing drift seedlings, black mangrove seedlings appear to be so delicate that they would not be able to withstand the vicissitudes of drifting or being stranded. Black mangrove disseminules are amazingly hardy, as Guppy (1917) discovered when he dried some mature fruits for 25 days at room temperature. The fruits lost 50 percent of their weight during the drying process. Yet they germinated when placed in fresh water. In our buoyancy tests, the fruits and seedlings often became soft and rotted, indicating that not all of them are as seaworthy as disseminules protected by a bony fruit or seed wall. Other common names for the black mangrove include salt-bush, because salt crystals are often gathered from the leaves, and honey mangrove, because of the excellent honey made from the floral nectar.

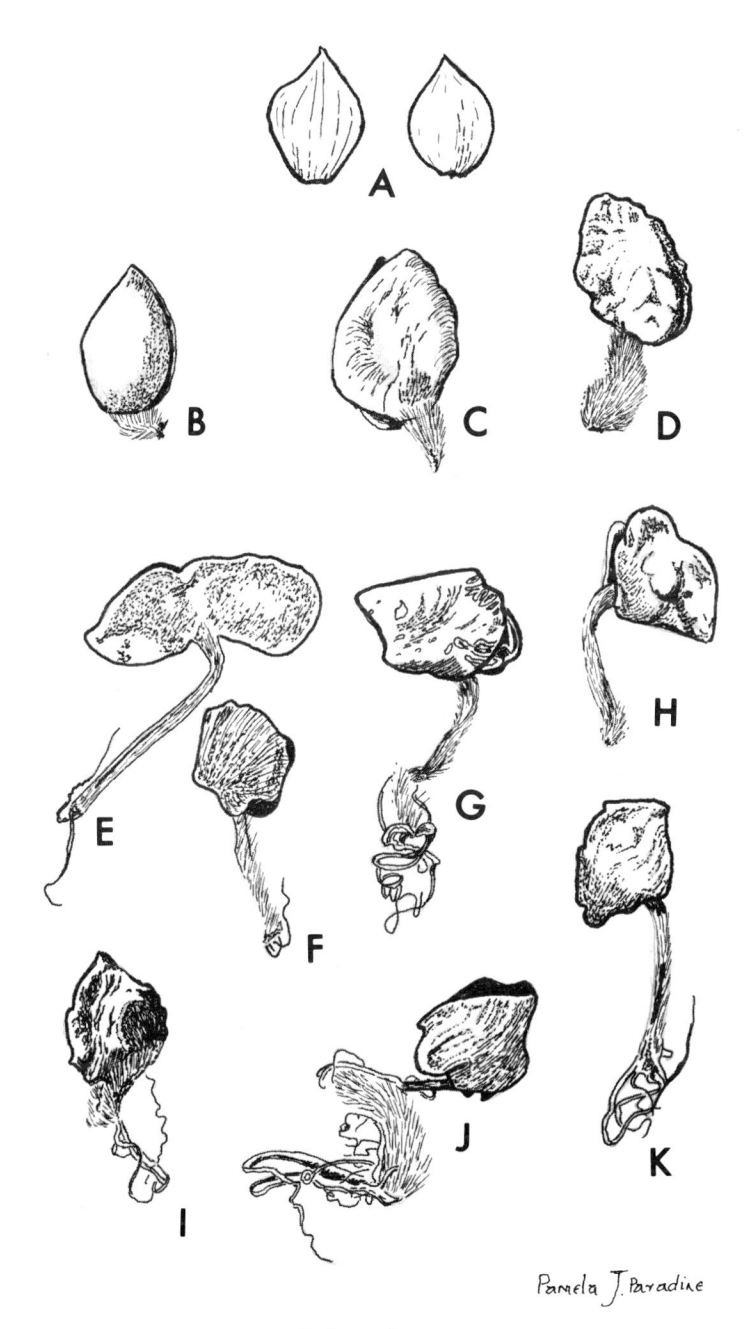

Avicenniaceae

AVICENNIA GERMINANS (L.) L.

Figure 23. *Avicennia germinans*. A, unsprouted fruits; B-D, sprouted fruits; E-K, seedlings (X1). Fruits and seedlings, Dennis, southeastern coast of Florida.

Barringtoniaceae

BARRINGTONIA ASIATICA (L.) KURZ, BOX FRUIT

Synonym: *B. speciosa* Forst. in most drift literature.
Disseminule: Fruit.
Description: Fruit (A-F) up to 12 cm long, 15 cm wide, top-shaped, 4- (rarely 6-) ridged at apex, tapering to base, square in cross section, shiny to dull gray, yellowish to dark brown, surface smooth to slightly wrinkled, if exocarp absent, surface fibrous.
Buoyancy factor: Fibrous mesocarp.
Buoyancy: At least 2 years.
Viability: None viable in our tests.
Currents: Most tropical currents.

Ideally suited for ocean dispersal, these fruits are among the largest, most durable, and widespread tropical drift disseminules. Because these fruits seldom contain a viable seed after drifting, sea currents have had little influence on plant dispersal, except locally. The fruits are readily identified by their box-like appearance, even though they vary considerably in size. Both large and small (5 cm long) fruits drift. The paper-thin exocarp is often partially or completely eroded revealing the thick fibrous mesocarp. In large fruits the mesocarp may be 2 cm thick (F), while in smaller ones it is about 1 cm thick. A single seed surrounded by a thin seed coat fills the seed cavity of fresh fruits. The embryo is unusual, because the cotyledons are undifferentiated and are seldom found in mature seeds. Grated seeds are used as fish poison. However, Arnold (1968) noted that the poisonous principle in the seeds seems to vary from place to place. Like coconuts (*Cocos nucifera*), box fruits may float for more than one year and take several years to disintegrate when stranded. Mesocarps make excellent fishnet floats. This species is pantropic, spreading from its southeast Asian home by man's activities. Unlike many other littoral trees which contribute long range drift fruits, the box fruit tree is usually found in a sand habitat, rather than in a mud habitat. Like the coconut, box fruit disseminules are often among the first arboreal disseminules to reach a newly formed tropical island. Treub (1888) found ten species of tropical fruits stranded on Krakatau after most of its vegetation was destroyed by volcanic eruptions. The box fruit was among these disseminules. For a discussion of Krakatau, see the Indian Ocean section of Chapter Three.

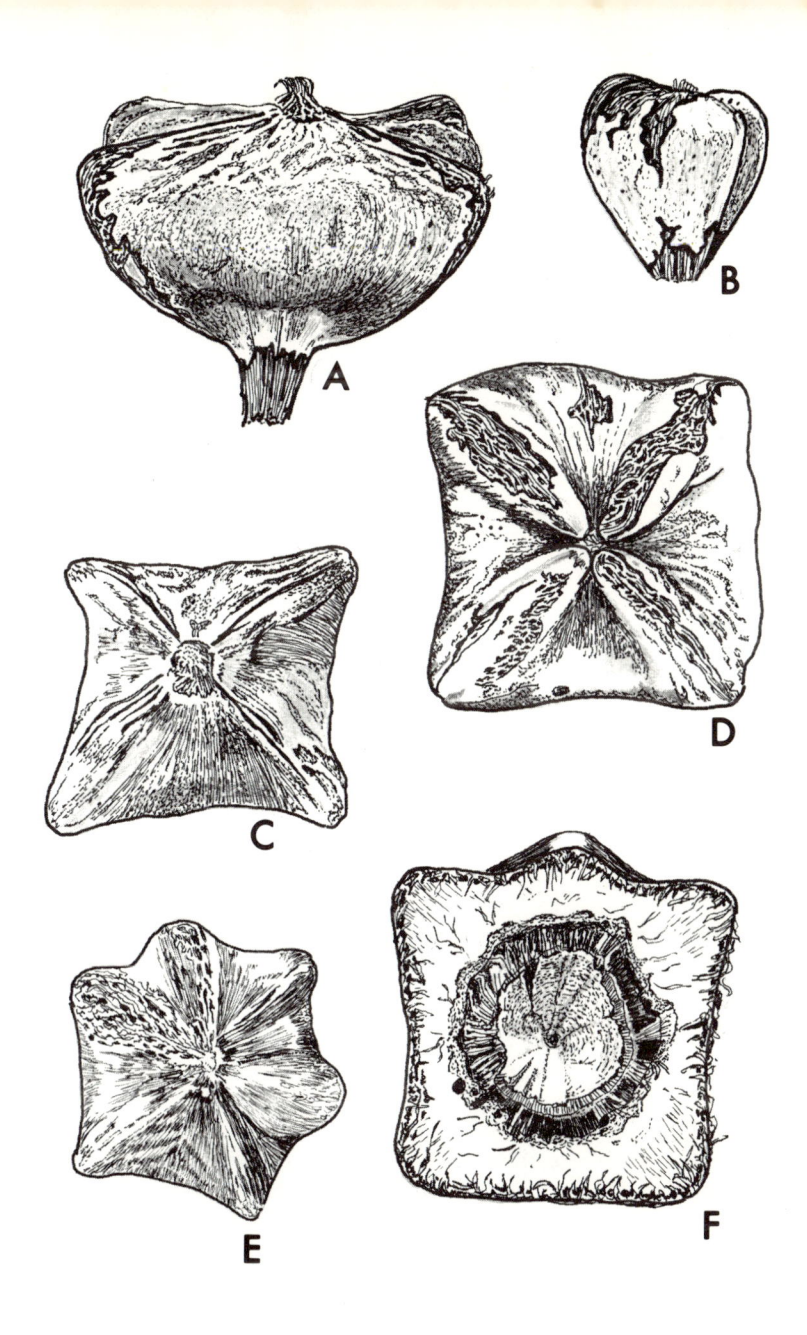

Barringtoniaceae

BARRINGTONIA ASIATICA (L.) KURZ

Figure 24. *Barringtonia asiatica* fruits. A, lateral view; B, miniature fruit in lateral view; C-E, apical views; F, cross section (X½). Fruits, Dennis, southeastern coast of Florida, and Degener and Degener, beaches of Canton Island.

Bignoniaceae

CRESCENTIA CUJETE L., CALABASH
DENDROSICUS LATIFOLIUS (MILLER) A. GENTRY

Desseminule: Fruit.
Description: *C. cujete* fruit (A-C) 6 to 16 cm long, 5 to 15 cm in diameter, globose to ellipsoidal, round in cross section, tan to nearly black, surface essentially smooth though numerous minute pits may be present, bearing a raised basal scar, shell hard and about 2 mm thick. *D. latifolius* fruit (D) 8 to 10 cm long, 5 to 8 cm in diameter, ellipsoidal, round in cross section, tan to nearly black, surface as above, bearing a flush basal scar, shell hard and about 1 mm thick.
Buoyancy factor: Fresh fruit—spongy tissue; old fruit—hollow interior.
Buoyancy: *C. cujete*—more than 1.5 years; *D. latifolius*—not tested.
Viability: *C. cujete*—no seeds viable; *D. latifolius*—not tested.
Currents: Tropical currents arising in the New World.

Fruits from these two native trees of the New World tropics illustrate the importance of the condition of the drifting disseminule. Fruits of *C. cujete* regularly reach the Carolina beaches and some have reached northern Europe. On the other hand, *D. latifolius*, formerly *C. cucurbitiana* L., fruits rarely reach Florida beaches. We believe that the main reason for this is that the fruit shell of *D. latifolius* is only 1 mm thick and often damaged at sea, while the fruit shell of *C. cujete* is 2 mm thick and more seaworthy. Old European drift literature (Guppy, 1917; Gunnerus, 1765; Lindman, 1882; Linnaeus, 1789; Sernander, 1901; and Ström, 1762) described drift gourds (*Lagenaria siceraria* (Mol.) Standley) and drift passionflower (*Passiflora*) fruits. We regard these as descriptions of *C. cujete* fruits which closely resemble gourd and passionflower fruits. Guppy (1917) separated the fruits of the two species, in part, by the presence of numerous minute pits on the surface of the fruit shell of *D. latifolius*. He stated that these pits were absent on fruit shells of *C. cujete*. We have found that this is not true. The pits occur on both fruits. Even after the fruits are thoroughly dried, they often emit a malodorous aroma. Fruits should be stored in a container or polyethylene bag. Calabash fruits are often used to make bowls or bailers.

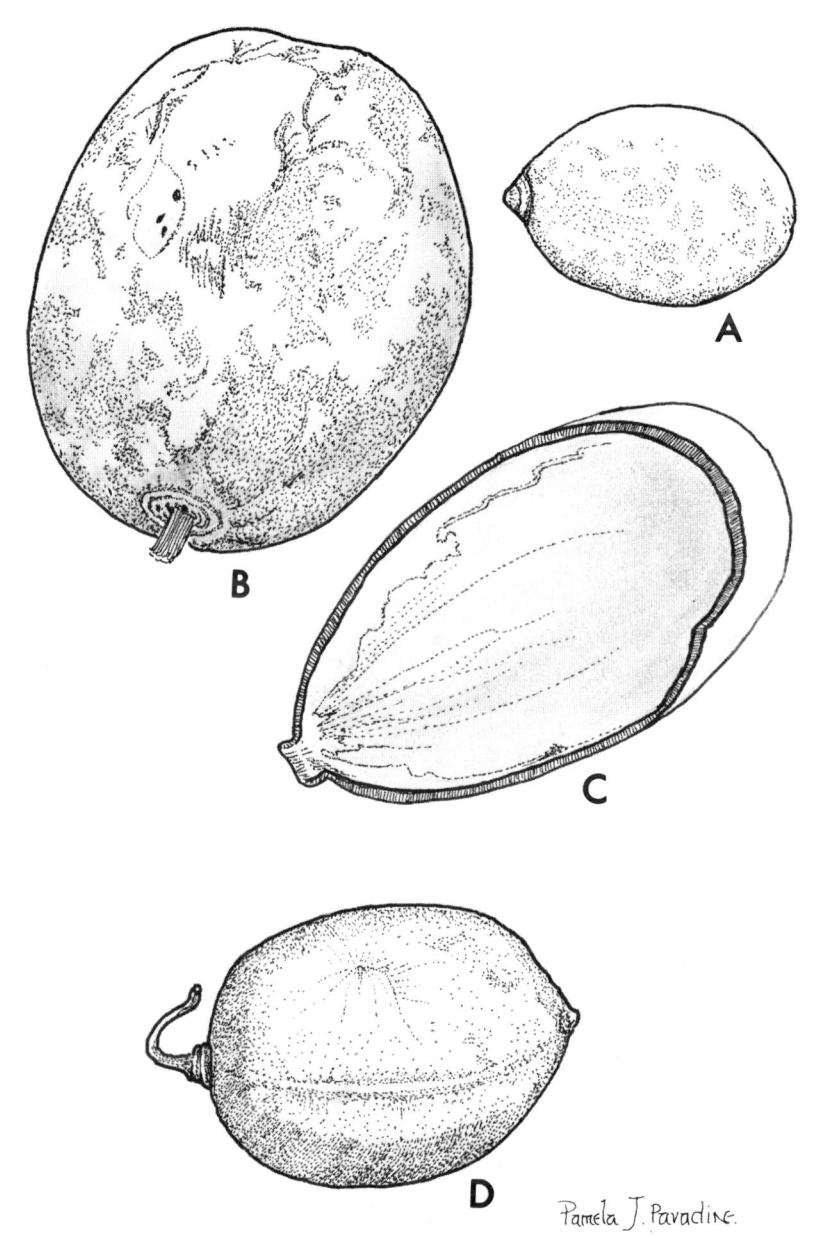

Bignoniaceae

CRESCENTIA CUJETE L., CALABASH
DENDROSICUS LATIFOLIUS (MILLER) A. GENTRY

Figure 25. *C. cujete* (A-C); *D. latifolius* (D). A-B, D, lateral views of fruits; C, longitudinal section (X½). Fruits, Gunn and Dennis, southeastern coast of Florida.

Burseraceae

CANARIUM DECUMANUM GAERTNER
CANARIUM MEHENBETHUNE GAERTNER

Disseminule: Endocarp.

Description: *C. decumanum* endocarp (A-B) 6 to 7 cm long, 4 to 6 cm wide, elliptical, triangular in cross section, brown, surface rugose with 3 apical ribs about half the length of the endocarp. *C. mehenbethune* endocarp (C-D) 4 to 7 cm long, 4 to 5 cm wide, globose, round in cross section, brown to blackish, surface smooth, apical ribs similar though only one quarter of the length of the endocarp. *C.* sp. endocarp (E-F) 4 to 6 cm long, 2.5 to 3 cm wide, elliptical, triangular in cross section, brown, surface minutely pitted, apical ribs absent.

Buoyancy factor: Empty or partially filled seed cavities.
Buoyancy: Not tested.
Viability: Not tested.
Currents: Tropical currents arising in the southeast Asia and Pacific regions.

Our identifications are based on illustrations in Schimper (1891, plate 6, figure 7b) and Gaertner (1789–1790, plate 102). The bony endocarps are 4 to 5 mm thick, and they are divided into 2 or 3 chambers. While the endocarp walls are strong enough to provide excellent protection for the seeds, their apices (especially the ribs) are weak. This defeats the value of the strong walls. While dead endocarps may be long range drifters, we doubt that drifting is an important mechanism for long range distribution. Most of the 100 species of *Canarium* in the Old World tropics are large trees of the rain forest. Several species bear edible seeds (Java almond and pili nut) and these may have been intentionally distributed by man.

Burseraceae

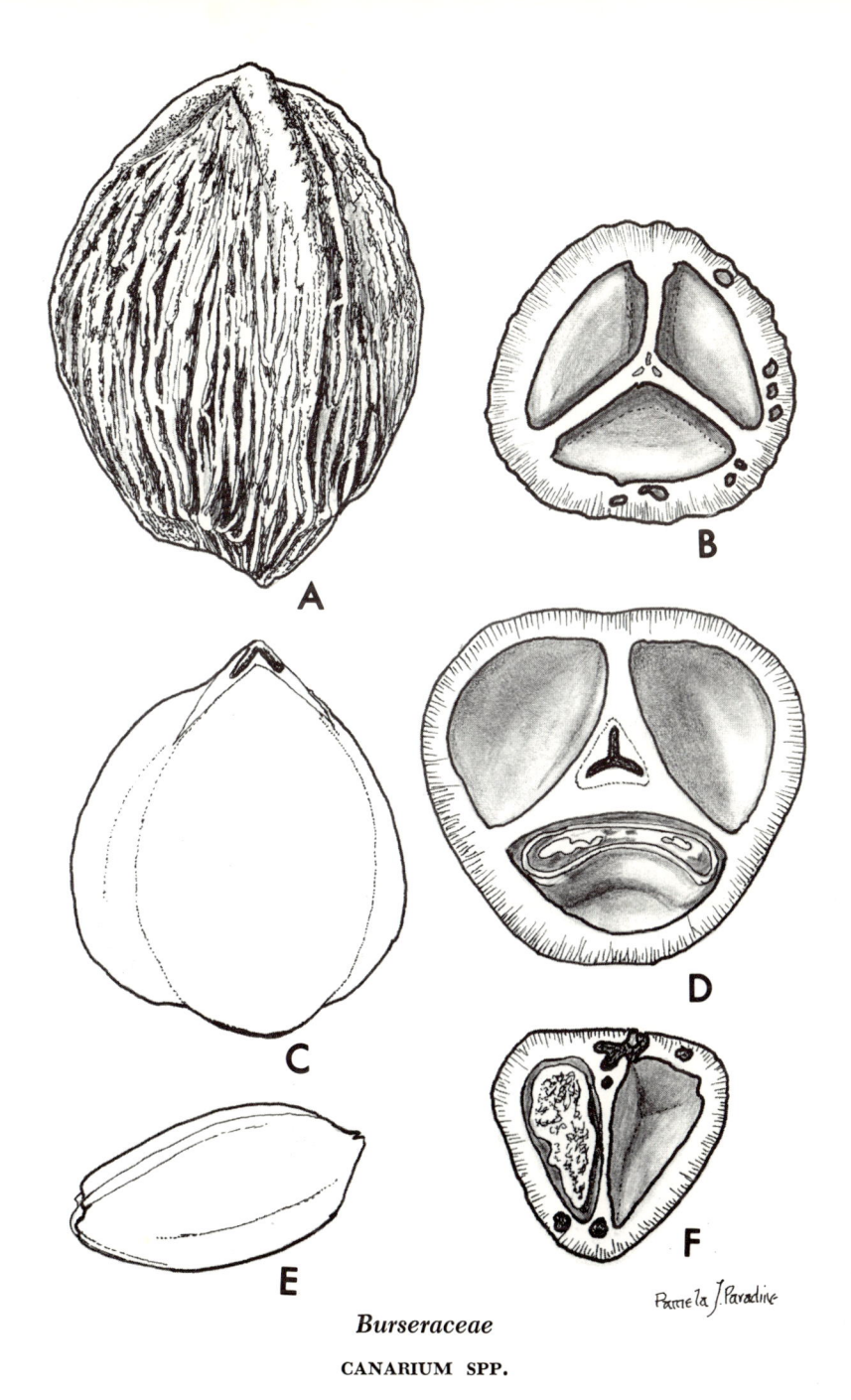

CANARIUM SPP.

Figure 26. *Canarium decumanum* (A-B); *C. mehenbethune* (C-D); *C.* sp. (E-F). A-F, endocarps. A, C, E, lateral views; B, D, F, cross sections (X1). Endocarps, Degener and Degener, beaches of Canton Island.

(85)

Caryocaraceae

CARYOCAR GLABRUM (AUBLET) PERSOON
CARYOCAR MICROCARPUM DUCKE

Disseminule: Endocarp, rarely eroded fruit.
Description: *C. glabrum* endocarp (B-D) 3 to 4 cm long, 2 to 3 cm wide, ellipsoidal with one side notched, ellipsoidal in cross section, blackish brown, surface covered by deeply sculptured (7 to 8 mm tall) flat-topped knobs. Fruit (A) about same size, globose, tan, surface slightly wrinkled. *C. microcarpum* endocarp (F-H) 2 to 4 cm long, 2.5 to 3 cm wide, ellipsoidal to C-shaped, compressed in cross section, blackish brown, surface covered by deeply sculptured (2 to 4 mm tall) spines. Fruit (E) about same size, ellipsoidal, tan, surface slightly wrinkled.
Buoyancy factor: Empty space around seed, seed absent, or perhaps seed bearing buoyant tissues.
Buoyancy: *C. glabrum*—more than 6 months; *C. microcarpum*—about 21 months.
Viability: *C. glabrum*—not tested; *C. microcarpum*—none viable.
Currents: Tropical currents arising in the New World.

Our disseminules were identified by G. T. Prance, New York Botanical Garden, who recently monographed the family (Prance and Freitas da Silva, 1973). The disseminules are produced by trees growing in periodically flooded forests throughout the Amazon-Orinoco River basin of northern South America. While these trees are widespread in northern South America, they have not spread to other tropical regions of the New World. Yet stranded disseminules have been collected on beaches as far north as the Carolinas (Gunn and Dennis, 1972b). Either the seeds are short lived, or they never reach a suitable habitat. The former seems more likely than the latter. Prance and Freitas da Silva reported an unusual collection of a *C. villosum* (Aublet) Persoon fruit from the shore of Barra in the Hebrides. The collection was made by Miss Somerville in 1959, and the fruit is deposited in the Royal Botanic Gardens Herbarium, Kew. This discovery is unusual for several reasons. The fruit does not appear to be seaworthy. No disseminule of this species has ever been recorded from any beach between Barra and northern South America. Finally, the trees inhabit the dry forest, not the flooded areas that the two other species inhabit.

Caryocaraceae

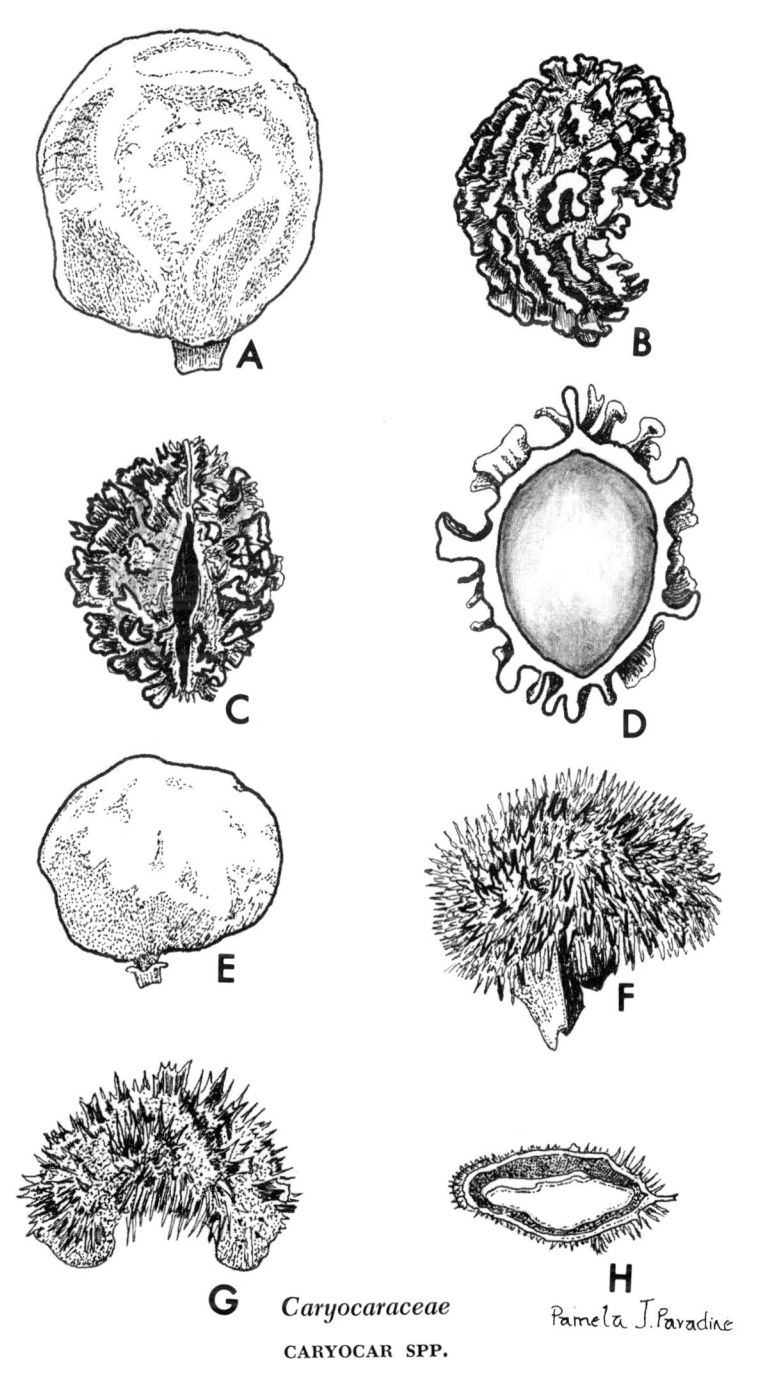

CARYOCAR SPP.

Figure 27. *C. glabrum* (A-D); *C. microcarpum* (E-H). A, E, fruits; B-D, F-H, endocarps. A-B, E-G, lateral views; C, suture view; D, H, cross sections (X1). Fruits, from trees; endocarps, Gunn and Dennis, southeastern coast of Florida.

Combretaceae

TERMINALIA CATAPPA L., COUNTRY-ALMOND
TERMINALIA SPP.

Disseminule: Mesocarp, rarely fruit.
Description: *T. catappa* mesocarp (Fig. 28 B-F) up to 7 cm long and 5 cm wide ellipsoidal, compressed in cross section, tan to straw colored, surface ribbed or grooved, pitted. Fruit (A) blackish brown, wrinkled. Other species are described below.
Buoyancy factor: Corky mesocarp.
Buoyancy: *T. catappa*—at least 2 years.
Viability: *T. catappa*—about 50 percent viable.
Currents: *T. catappa*—most tropical currents; others—as noted below.

Country-almond, also known as tropical- or Indian-almond, is one of the most common tropical drift fruits. Because the mesocarp is rather soft (thus quite buoyant), it is found in various states of erosion. Occasionally intact fruits are stranded. The species, a native of tropical Asia, has been intentionally spread by man throughout the tropics. No doubt tropical currents and bats have contributed to this distribution. Besides producing edible seeds, which taste like almonds, the trees are an excellent shade tree. Fruits vary in size depending on whether the fruits came from wild or naturalized trees, or from cultivated trees which have been selected for seed size. The country-almond is the only tropical disseminule which has been recorded to have been stranded on the Oregon coast (Gibbons, 1967). The disseminules must have drifted from the Kuroshio Current into the North Pacific Current, and then into the California Current. There is no current which flows south to north along the west coast of the United States.

Other species of *Terminalia* have drift disseminules. Except for C and G, the disseminules shown in Figure 29 more or less resembles country-almond mesocarps. All of these disseminules were collected on the beaches of Canton Island. The disseminules shown in Figure 30 came from these areas: A, Pacific region; B-D, Caribbean region. Only B was collected from a beach. The others float but were collected from the area of the parent plant.

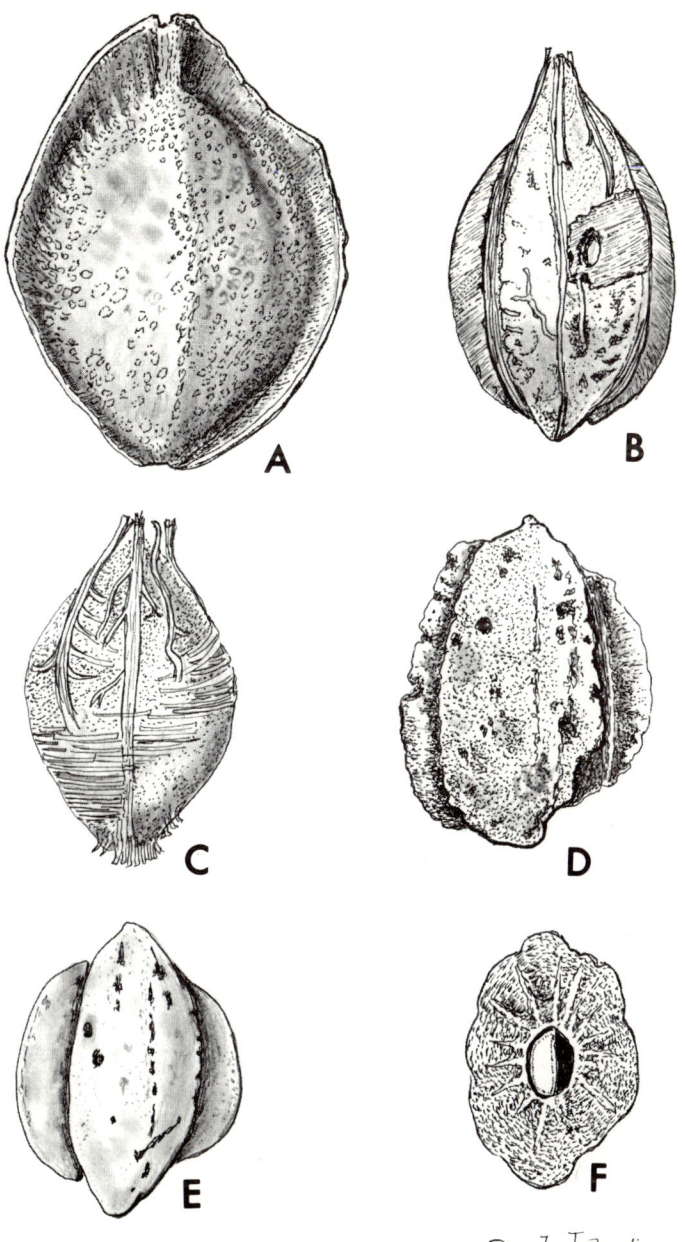

Combretaceae

TERMINALIA CATAPPA L.

Figure 28. *Terminalia catappa*. A, fruit; B-F, mesocarps. A-E, lateral views; F, cross section (X1). Fruit and mesocarps, Gunn and Dennis, southeastern coast of Florida.

(89)

Combretaceae

TERMINALIA SPP.

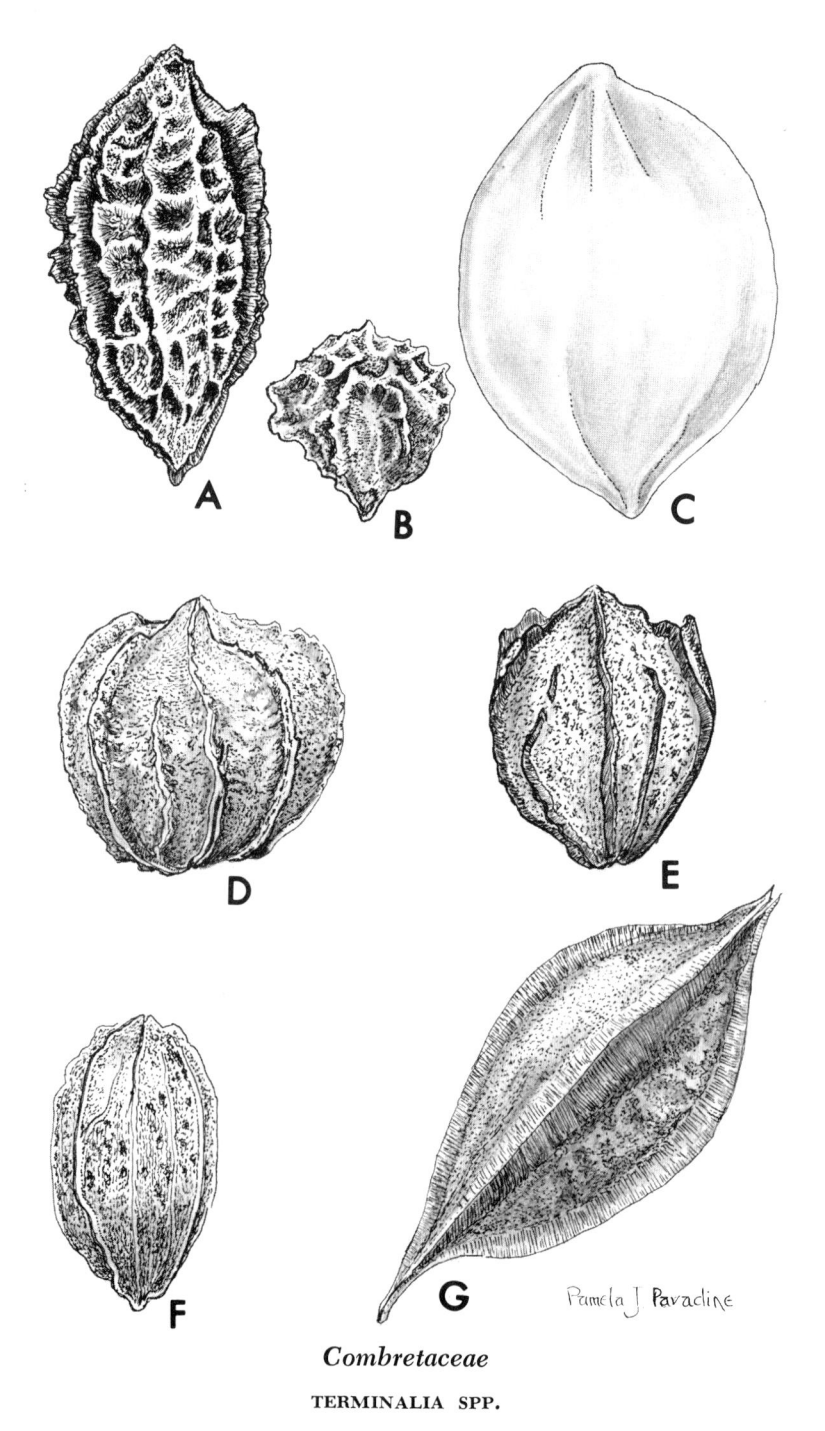

Figure 29. *Terminalia* spp. mesocarps. A-G, lateral views (X1). Mesocarps, Degener and Degener, beaches of Canton Island.

(90)

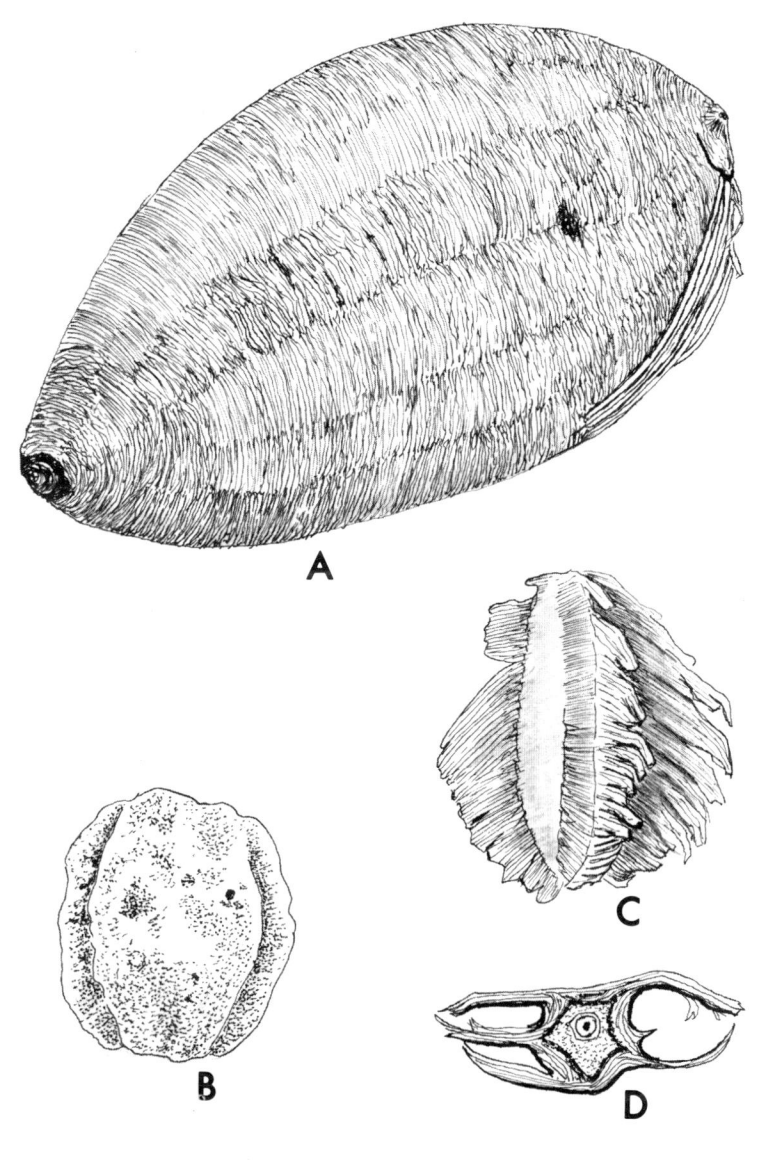

Combretaceae

TERMINALIA SPP.

Figure 30. *Terminalia* spp. A, C-D, fruits, B, mesocarp. A-C, lateral views; D, cross section of C (X1). Fruits and mesocarp from trees.

Convolvulaceae

IPOMOEA ALBA L., WHITE MOONFLOWER
IPOMOEA MACRANTHA R. & S., SEA MOONFLOWER
IPOMOEA PES-CAPRAE (L.) R. BR., RAILROADVINE

Disseminule: Seed.
Buoyancy: Cavity within seed.
For other data see the species discussions and Gunn (1969, 1972a).

Ipomoea alba seed (A) 10 to 12 mm long, 8 to 9 mm wide, spheroid, round in cross section, monochrome ochre, brown, or blackish brown, surface smooth. White moonflower is a perennial pantropic vine which has been used ornamentally for its spectacular flowers both in the tropics and in the temperate zone. In the drift literature, the white moonflower is placed in the segregate genus *Calonyction*. Muir (1930) germinated drift seeds from his Riversdale, South Africa site. He also found that of the 70 seeds initially buoyant, 59 floated for 149 days. In our tests seeds floated for more than 1.5 years. Seeds may be found in most tropical currents.

Ipomoea macrantha seed (B) 10 to 11 mm long, 8 to 9 mm wide, spheroid, triangular in cross section, dark brown, surface covered with minute pale hairs with longer hairs restricted to the margin and hilum. The sea moonflower, known as *I. tuba* (Schlect.) G. Don or *Calonyction tuba* (Schlect.) Colla in the drift literature, is a pantropic vine of the back beach flora. Guppy (1917) reported that seeds remained afloat for ten weeks, and they appeared to be able to float much longer. Seeds may be found in most tropical currents.

Ipomoea pes-caprae seed (C) 6.5 to 7 cm long and wide, wedge-shaped, triangular in cross section, medium brown, surface densely covered with short pale hairs. Railroadvine is a ubiquitous pantropic beach vine. It forms great carpets of vegetation just beyond the high tide zone. Seeds float for at least 2 years, and they may be expected in any tropical current. Railroadvine has become established on Ninety Mile Beach, North Auckland, New Zealand by drifting to the beach in recent years (Cooper, 1967; Mason, 1961; Sykes, 1970).

Ipomoea spp. Seeds of these species of *Ipomoea* float and are similar in appearance to the three described species. Figure 31 D depicts seeds similar to *I. alba* which were stranded on beaches of southeastern Florida. These seeds are less than 9 mm long or less than 7 mm wide and are reddish brown to tan. Other seeds (E-F) have also been collected from southeastern Florida beaches.

Convolvulaceae

IPOMOEA SPP.

Figure 31. *Ipomoea alba* (A); *I. macrantha* (B); *I. pes-caprae* (C); *I.* spp. (D-F). A-F, seeds (X2). Seeds, Gunn and Dennis, southeastern coast of Florida.

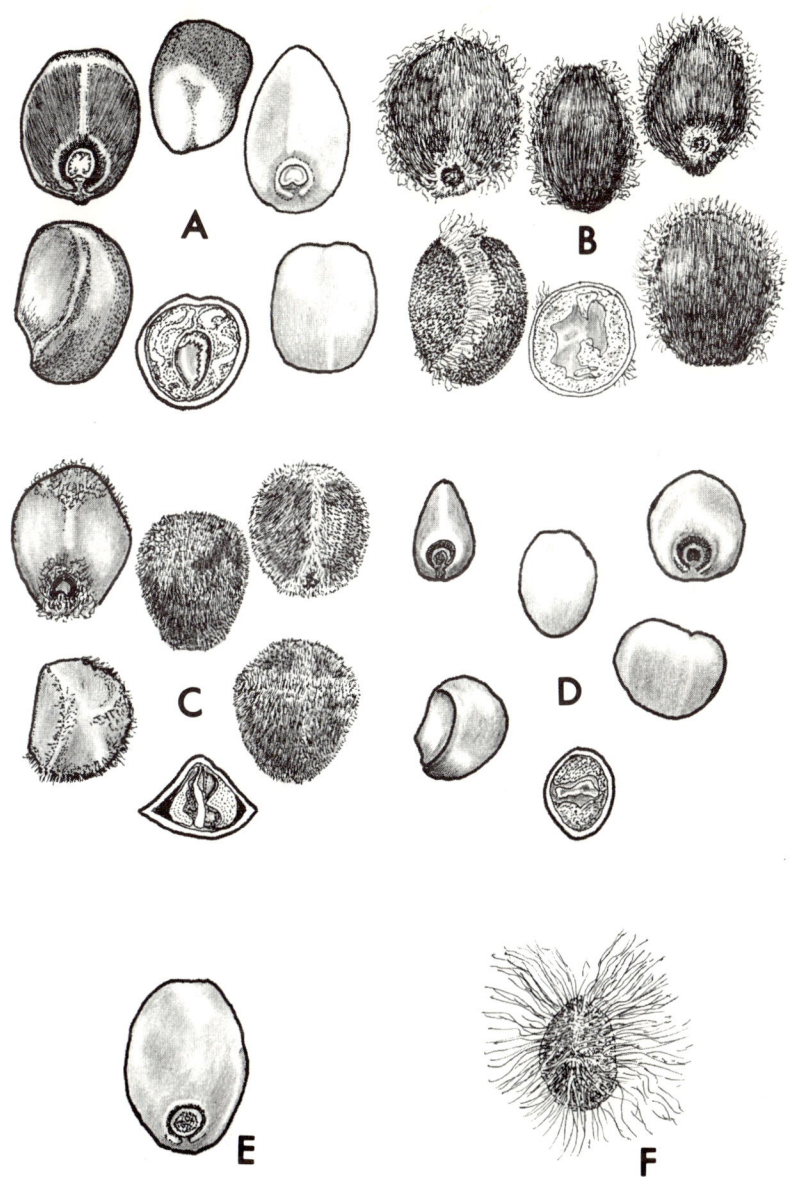

Convolvulaceae

MERREMIA DISCOIDESPERMA (DONN. SM.) O'DONELL, MARY'S-BEAN

MERREMIA TUBEROSA (L.) RENDLE, WOOD-ROSE

Synonyms: *Merremia* is *Ipomoea* in drift literature; *M. discoidesperma* is erroneously named *I. tuberosa* L. in some drift literature.

Disseminule: Seed.

Description: *M. discoidesperma* seed (A-F) 20 to 30 mm in diameter, 15 to 20 mm thick, globose to oblong, compressed in cross section, lustrous to dull, black to brown or combination of two colors, surface smooth except for two grooves which form a cross (C-D). Hilum large, seal-like, often light colored (A, B, E). *M. tuberosa* seed (G-I) 10 to 25 mm in diameter and thickness, subglobose, rounded to triangular in cross section, lustrous to dull blackish brown, smooth. Hilum large, seal-like, often light colored.

Buoyancy factor: Cavity within seed.

Buoyancy: *M. discoidesperma*—at least 2 years; *M. tuberosa*—not tested.

Viability: *M. discoidesperma*—most viable; *M. tuberosa*—not tested.

Currents: Tropical currents arising in the New World; *M. tuberosa* may also be found in Old World currents.

No drift seed has had a longer and more intriguing history than Mary's-bean. Historical notes are presented in Chapter 2. Few New World plants are as poorly collected or as little known as the Mary's-bean. This woody vine is a member of the Guatemalan wet or mixed forest flora. It has also been reported from Chiapas, Mexico and Hispaniola (Standley and Williams, 1970). Hemsley (1892) incorrectly identified the seeds of *M. discoidesperma* as *M. tuberosa*. This error has been perpetuated by others. Mary's-bean does not seem to spread by drifting even though the seeds may be viable when stranded. We have found that seedlings in Florida are quickly and completely defoliated by insects, and this might explain why the vine has not spread. The seeds are atypical, because one seed occupies the place of 4 in the fruit. Compare these seeds with *M. tuberosa* or *Ipomoea* seeds. Williams (1973) presents additional information, and Gunn (1976) documents the occurrence of drift seeds on Norwegian and Wotho atoll, Marshall Islands beaches and presents a taxonomic treatment.

Wood-rose seeds are found less frequently on New World beaches than Mary's-bean seeds. Little is known about their buoyancy or occurrence, because they have seldom been identified correctly. The fruits are the famous wood-roses of dried arrangements. The vine has attained a pantropic distribution, because of its ornamental value.

Convolvulaceae

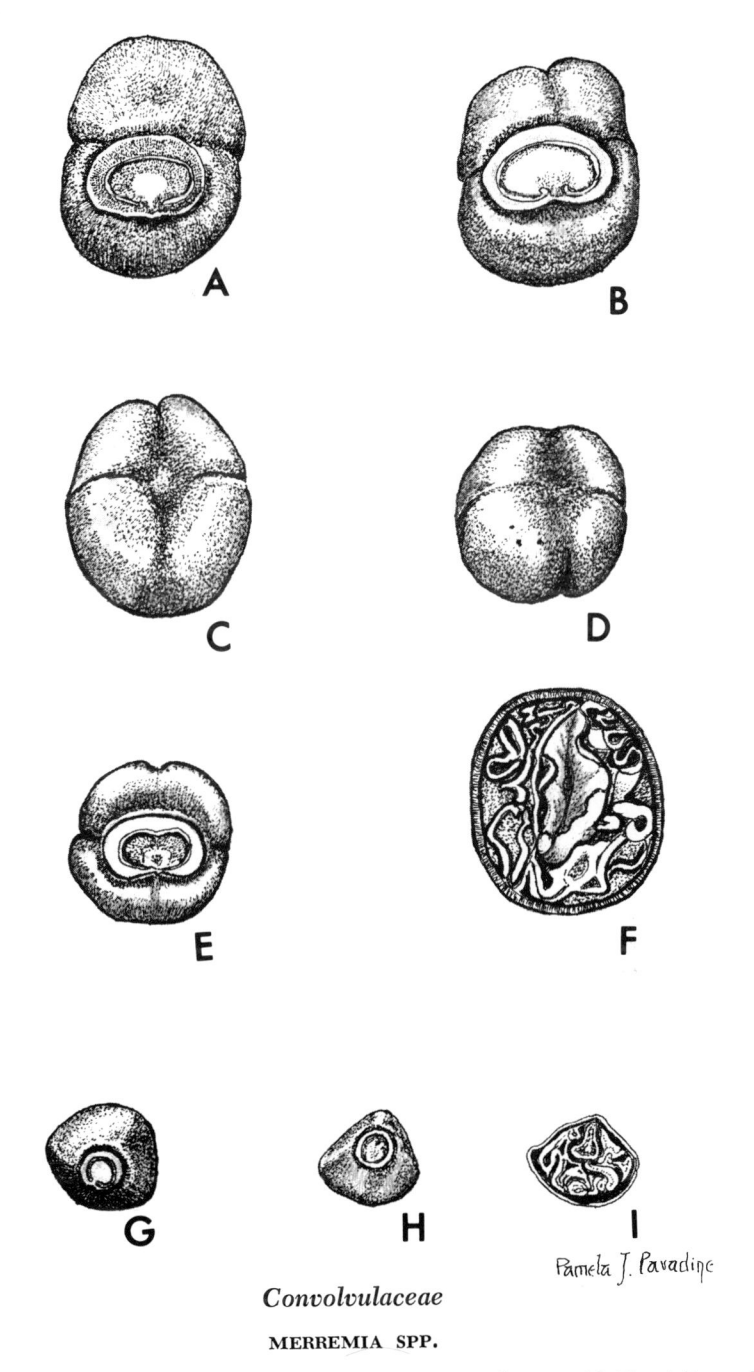

MERREMIA SPP.

Figure 32. *Merremia discoidesperma* (A-F); *M. tuberosa* (G-I). A-I, seeds. A-B, E, G-H, hilar views; C-D, dorsal views; F, I, cross sections (X1). Seeds, Gunn and Dennis, southeastern coast of Florida.

Cucurbitaceae

FEVILLEA CORDIFOLIA L., ANTIDOTE VINE

Disseminule: Seed or half seed coat.
Description: Seed (A-F) 5 to 6 cm in diameter, 2 cm thick, round to angular in outline, strongly compressed in cross section, ochre to dull sepia gray, outer surface minutely roughened and somewhat bent, inner surface slightly roughened.
Buoyancy factor: Corky seed coat and perhaps partially filled seed cavity.
Buoyancy: About 22 months.
Viability: About 20 percent viable.
Currents: Tropical currents arising in the New World.

Antidote vine or antidote caccoon is a high climbing woody vine which inhabits fresh water swamps in the New World tropics. Fruits are 10 to 12 cm in diameter, globose, russet, hard, and about 10-seeded. Fruits falling into fresh water float, but are usually eroded by the time they reach the ocean. Seeds once freed from the fruit often germinate in the fresh water. Seawater is lethal. The seedlings will float along as they are attached to the seed coat. Ungerminated seeds which reach the ocean may drift. While Guppy (1917) estimated that only 5 percent of the drift seeds were viable, we found about 20 percent of the undamaged drift seeds on Florida beaches were viable. Damaged seeds and half seed coats are often found stranded, indicating that the seeds are fragile, and that the corky seed coat is buoyant. We agree with Ridley (1930) that fresh water dispersal is more important than ocean current dispersal. Fresh seeds are a source of oil which burns when ignited.

Cucurbitaceae

FEVILLEA CORDIFOLIA L.

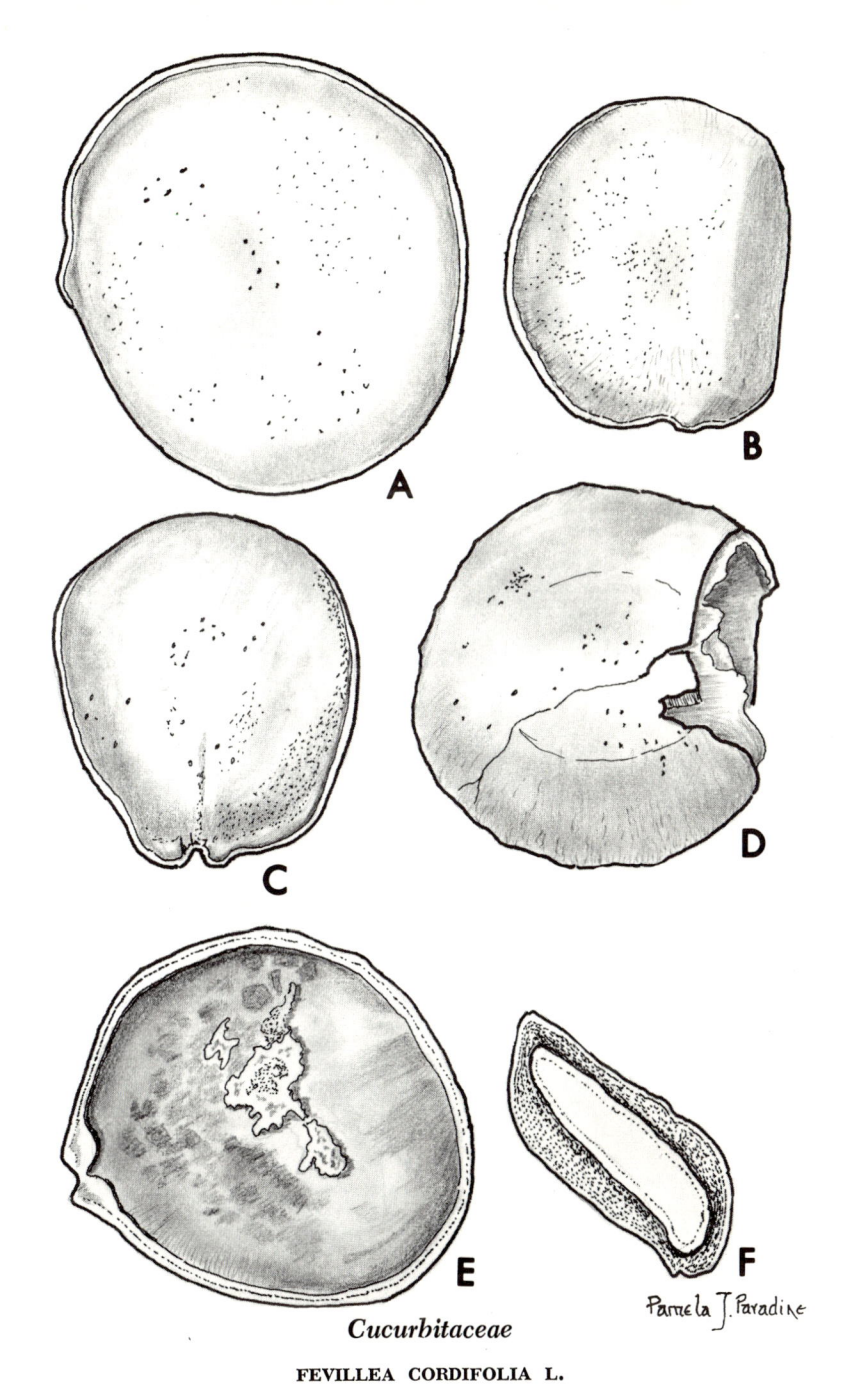

Figure 33. *Fevillea cordifolia* seeds. A-D, lateral views; E, inner face of seed coat; F, cross section (X1). Seeds, Gunn and Dennis, southeastern coast of Florida.

Cycadaceae

CYCAS CIRCINALIS L., FERN-PALM

Disseminule: Seed.

Description: Seed (A-D) 3 to 6 cm long, 3 to 4 cm in diameter, ellipsoidal to globose, apex often minutely short pointed, tapering to base, round in cross section, deep reddish brown to deep yellowish ochre, surface wrinkled to smooth.

Buoyancy factor: Corky layer in seed coat and empty space around embryo.

Buoyancy: About 3 months, though Ridley (1930) reported "some months."

Viability: Not tested.

Currents: Most tropical currents.

The fern-palm or queen sago is a widely planted ornamental tree which now has a pantropic distribution. Its natural range, thought to be the result of drifting, is from East Africa to Polynesia. Unlike seeds of the other drift species, the seeds of the fern-palm are not formed within a fruit. The outer layer of the 3-layered seed coat functions as a surrogate fruit. The outer layer (A) is fleshy and usually eroded on drift seeds. The hard middle layer (B-C), the surface seen most often on drift seeds, is yellowish ochre and about 1 mm thick. This layer serves the buoyant seed in the same fashion as the endocarp for a buoyant fruit. The inner layer is the buoyant layer, and it is brown, corky, and 1 to 5 mm thick. Fresh seeds are poisonous and can only be eaten (and then not safely as a major part of the diet) after soaking in several changes of fresh water, and then cooking or sun drying the seeds. The water used in soaking may be poisonous.

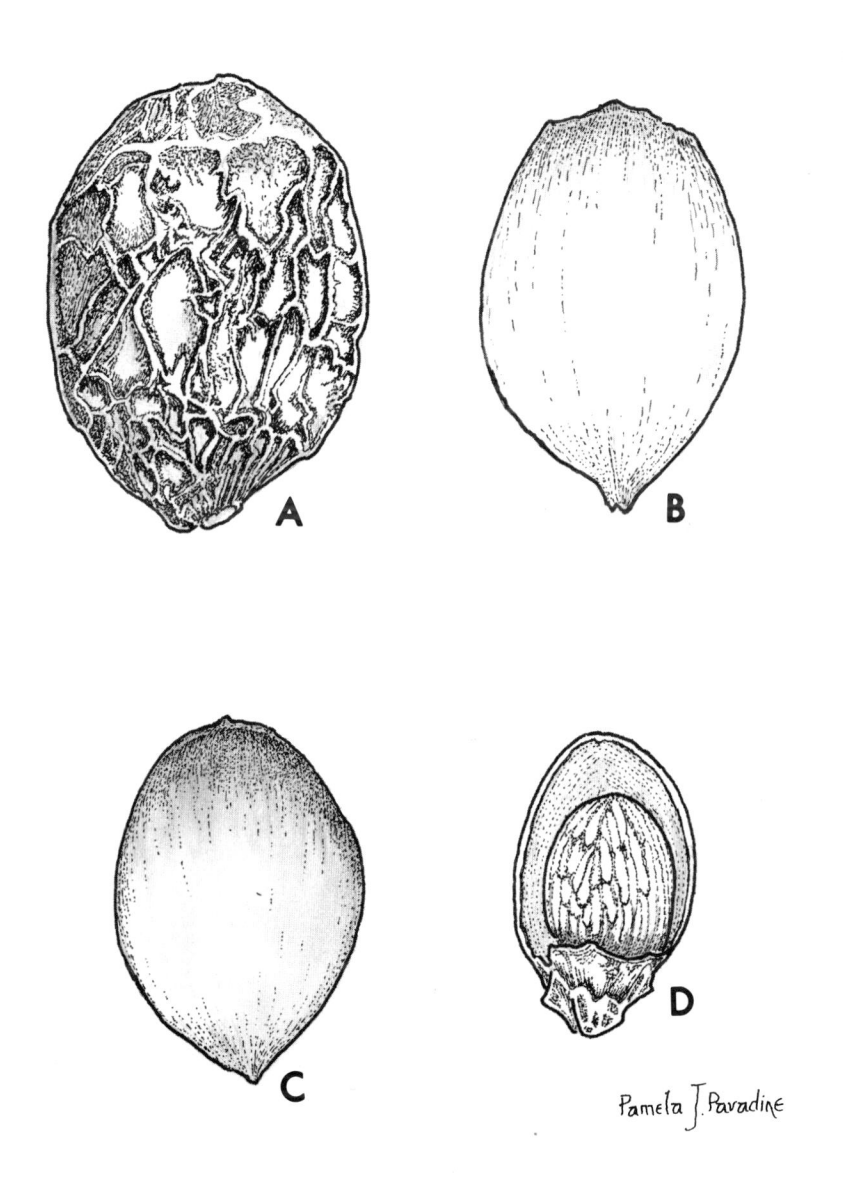

Cycadaceae

CYCAS CIRCINALIS L.

Figure 34. *Cycas circinalis* seeds. A-C, lateral views; D, longitudinal section (X_1). A, from tree. B-D, Degener and Degener, beaches of Canton Island.

Euphorbiaceae

ALEUTRITES MOLUCCANA (L.) WILLD., CANDLENUT

Synonym: *A. triloba* Forst. in Muir (1937).
Disseminule: Seed.
Description: Seed (A-F) 2.5 to 3.5 cm in diameter, subglobose, rounded at apex, pointed at base, slightly compressed in cross section, black to gray or brown, wrinkled to nearly smooth (E).
Buoyancy factor: Empty to nearly empty seeds.
Buoyancy: About 5 months.
Viability: Embryo absent or nearly so.
Currents: Most tropical currents.

The candlenut tree is often known as the Jamaican-walnut in the New World, because its seeds superficially resemble a walnut (*Juglans*) endocarp. In Florida, the drift seeds are appropriately called fossil-prunes. The tree, a native of Asia, has been spread by man throughout the tropics, because its seeds are rich in oil. Oil expressed from the seeds is used as a light source and as a mild cathartic. Degener (1945) gives an excellent account of the uses of the seeds in the Pacific area. Drift seeds seldom contain a viable embryo, thus distribution of the species cannot be attributed to ocean currents. Seeds containing sound embryos and fresh fruits may be washed down rivers and become stranded on the banks of estuaries where the seeds may germinate. No ocean transport is involved in this movement.

We have tentatively identified a few tung (*A. fordii* Hemsley) seeds from beaches of southeastern Florida. The *Aleurites*-like seeds (G-H) from the beaches of Canton Island resemble walnut (*Juglans*) endocarps more than the seeds of the candlenut. These Canton seeds often bear a better developed embryo than do the candlenut seeds.

Euphorbiaceae

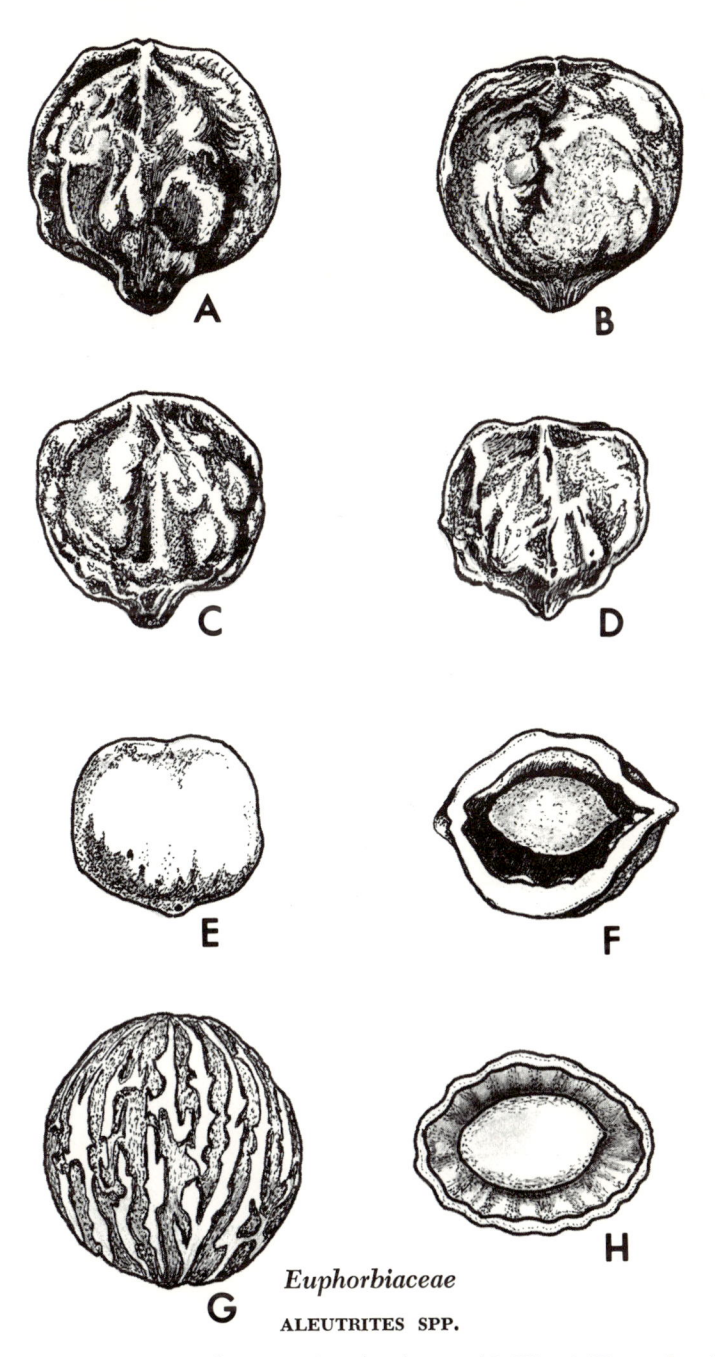

ALEUTRITES SPP.

Figure 35. *Aleurites moluccana* (A-F); *A.* sp. (G-H). A-H, seeds. A-E, G, lateral views; F, H, cross sections (X1). *A. moluccana*, Gunn and Dennis, southeastern coast of Florida and Degener and Degener, beaches of Canton Island; *A.* sp., Degener and Degener, beaches of Canton Island.

Euphorbiaceae

HIPPOMANE MANCINELLA L., MANCHINEEL

Disseminule: Mesocarp or endocarp.
Description: Mesocarp (A-F, J) 1.5 to 3.5 cm in diameter, compressed globose, round in cross section, light to dark tan, surface pitted, bumpy, often bearing a few to 12 grooves extending from base to apex. As the corky mesocarp erodes, the deeply sculptured bony endocarp (G-I) becomes exposed. The bumps or spines on the surface of the mesocarp are the tips of the endocarp rays.
Buoyancy factor: Corky mesocarp. Naked endocarps apparently roll in with the tide like shells.
Buoyancy: At least 2 years.
Viability: About 50 percent of the seeds are viable.
Currents: Tropical currents arising in the New World and perhaps in the Old World.

Manchineel trees form dense thickets along the coast of many tropical islands in the Caribbean Sea, as well as along the Atlantic and Pacific coasts of Central America and northern South America. The crabapple-like fruits and other parts of the tree are extremely poisonous. The poison is concentrated in the milky juice which is absent in drift disseminules. The juice will blister the skin, and smoke from burning wood will irritate eyes. Wood used as meat skewers will poison meat. The tree is described by Dahlgren and Standley (1944) and Johnston (1949).

Fresh fruits are buoyant. They are globular, yellow with reddish cheeks when ripe, and about 3 cm in diameter. The fleshy exocarp erodes quickly in ocean water exposing the mesocarp. As the mesocarp erodes, the deeply sculptured bony endocarp is exposed. These stellate endocarps are highly prized by collectors. Freed seeds are not buoyant. A. R. Melville regards K and L as manchineel mesocarps. These disseminules were collected from the beaches of Canton Island. We find too many discrepancies between manchineel endocarps and the specimens illustrated in K and L to label them *H. mancinella*. They are being shown here, because they resemble manchineel.

Euphorbiaceae

HIPPOMANE SPP.

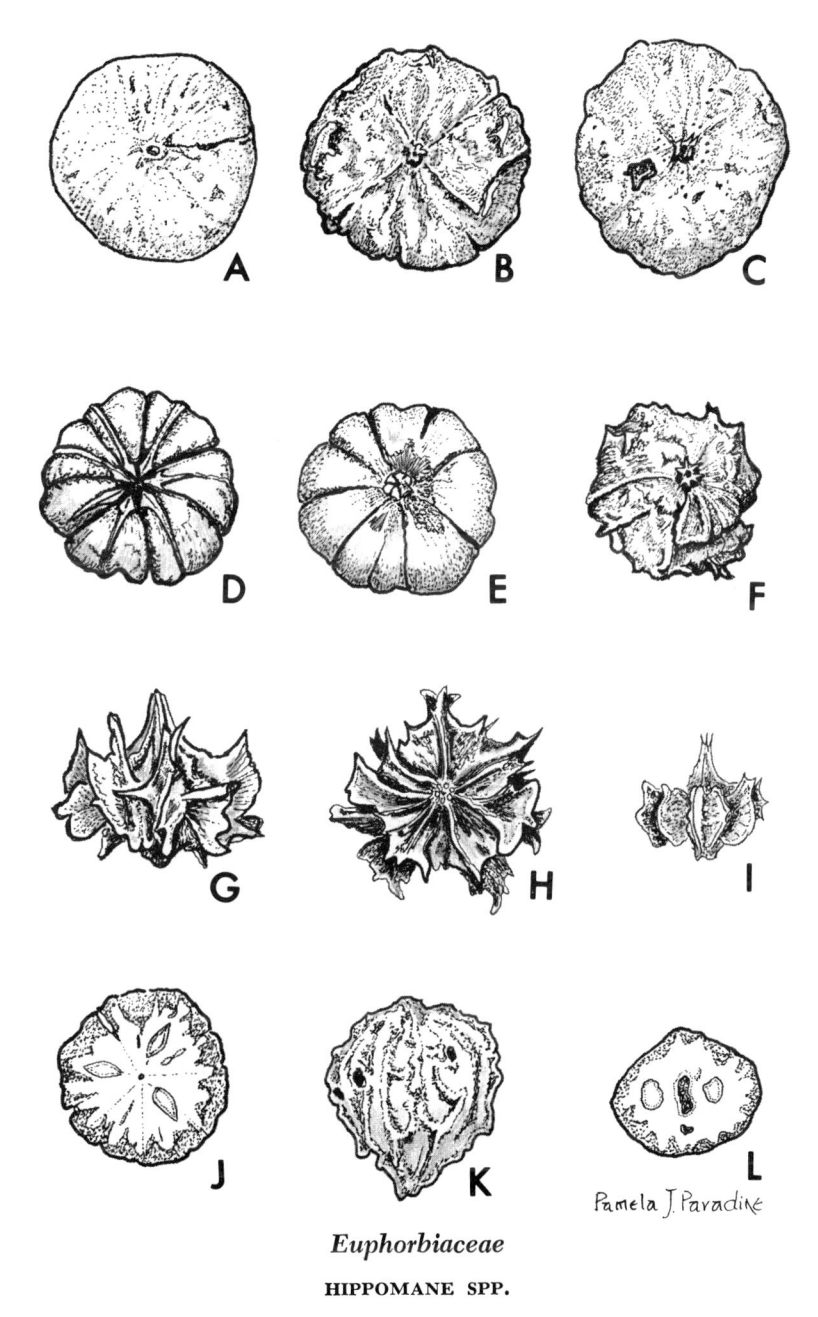

Figure 36. *Hippomane mancinella* (A-J); *H.* sp. ? (K-L). A-F, J-L, mesocarps; G-I, endocarps. A-F, H, basal views; G, I, K, lateral views; J, L, cross sections (X1). A-J, Gunn and Dennis, southeastern coast of Florida; K-L, Degener and Degener, beaches of Canton Island.

Euphorbiaceae

OMPHALEA DIANDRA L., JAMAICAN NAVEL SPURGE

Disseminule: Seed.

Description: Seed (A-F) 4 to 5 cm in diameter, ovate to circular, slightly compressed in cross section, upper side convex, lower side obliquely keeled, dark brown to nearly black or grayish black, surface finely tuberculate.

Buoyancy factor: Intercotyledonary cavity.

Buoyancy: At least 2 years.

Viability: About half of the unbroken seeds are viable.

Currents: Tropical currents arising in the New World.

Jamaican navel spurge, a trailing or climbing shrub of the New World tropics, often grows in thickets near beaches, tidal swamps, or rivers. Seeds, tightly packed in orange-size hard-shelled fruits, are freed when the fruits are ruptured. Seeds which reach the ocean may drift. Because the seed coat is brittle, seeds are often stranded in a damaged condition (C, E). These damaged seeds have lost their ability to germinate. Guppy (1917) reported that about half of the seeds he collected from West Indian beaches were empty. We have found that most seeds collected from United States beaches bear an embryo. Johnston (1949) reported seeds of *O. panamensis* (Beurl.) I. M. Johnston among the stranded disseminules of San Jose Island. Some botanists regard this species as a variety of *O. diandra*. The seeds which Johnston photographed closely resemble those of *O. diandra*. A related species, *O. triandra* L. (G), produces buoyant seeds that bear a softer seed coat. Thus, these seeds are less likely to drift for any appreciable distance.

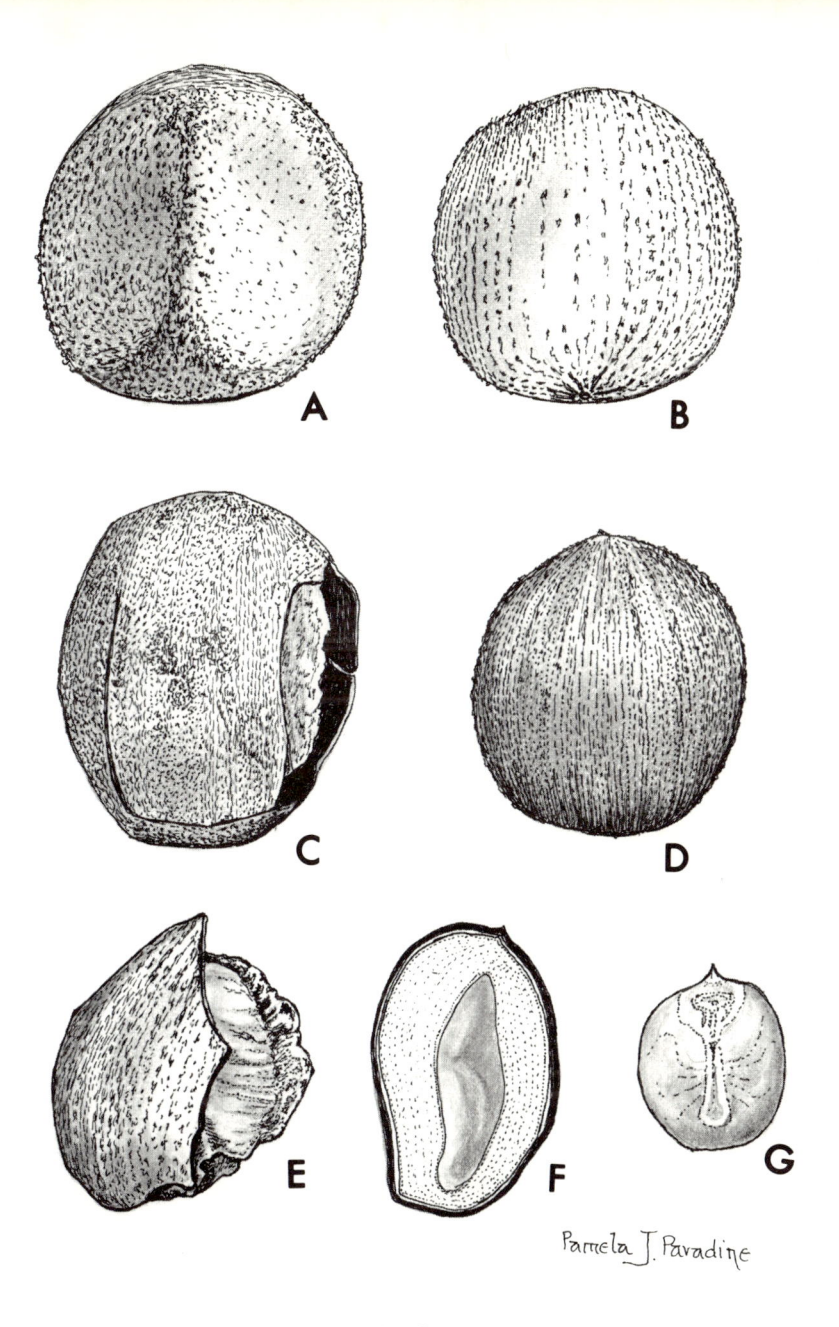

Euphorbiaceae

OMPHALEA SPP.

Figure 37. *Omphalea diandra* (A-F); *O. triandra* (G). A-G, seeds. A-B, D, G, lateral views of entire seeds; C, E, lateral views of broken seeds; F, cross section (X1). A-F, Gunn and Dennis, southeastern coast of Florida; G, from plant.

Fagaceae

QUERCUS BENNETTII MIQ.
QUERCUS SPP., OAKS

Disseminule: Fruit with or without cupule.

Description: *Q. bennettii fruit* (A-D) 2 to 3 cm long, 1.5 to 2.5 cm in diameter, elongate to subglobose, rounded at apex and truncate at base, round in cross section, grayish brown to grayish black, surface smooth and bearing a minute apical scar or spine and a large circular basal scar (C). *Q.* spp. fruit (E-I) are similar to the foregoing except in size. They are usually smaller than *Q. bennettii*.

Buoyancy factor: Space around embryo.

Buoyancy: Not tested.

Viability: Not tested. Most *Q.* spp. were damaged and thought to be dead.

Currents: *Q. bennettii*—tropical currents in the Pacific region; *Q.* spp.—at least tropical currents arising in the New World.

There is little known about the drift capacity of the acorns (fruits) of *Q. bennettii*. While this species is not widely planted in the New World tropics, it is well established in the Philippines and adjacent Pacific region.

Oaks are mainly temperate trees. They may be members of the inland forest, or of the back beach vegetation. Their acorns may be carried by rivers to the ocean where some may drift. Eastern and southern United States beaches receive acorns from several species. These acorns are not thought to drift far and often bear dead seeds when stranded (Guppy, 1917; Ridley, 1930). Dispersal by rivers is more important than ocean drift.

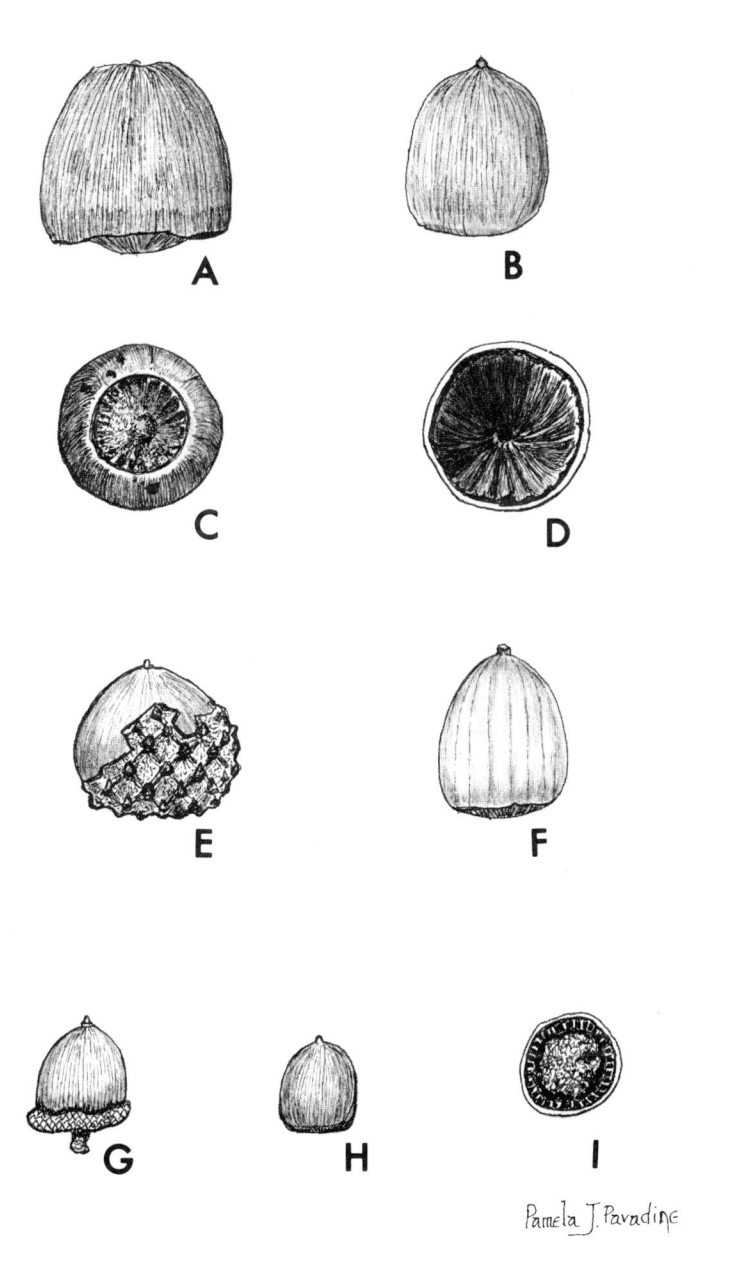

Fagaceae

QUERCUS SPP.

Figure 38. *Quercus bennettii* (A-D); *Q.* spp. (E-I). A-I, fruits. A-B, E-H, lateral views; C, basal view; D, I, cross section; G, cupule partially present (X1). A-D, Degener and Degener, beaches of Canton Island; E-I, Gunn and Dennis, southeastern coast of Florida.

(107)

Flacourtiaceae

PANGIUM EDULE REINW.

Disseminule: Seed.
Description: Seed (A-F) 3 to 6 cm long, 2.5 to 4 cm wide, round to triangular, compressed in cross section, gray or black, surface bearing conspicuous branched grooves. Hilum 4 to 6 cm long, prominent, resembling lips (E).
Buoyant factor: Empty seed.
Buoyancy: At least 2 years.
Viability: Embryo absent.
Currents: Tropical currents arising in the southeastern Asia and Pacific regions, and rarely in tropical currents of the New World.

This tree often grows along rivers, and its seeds are carried to the ocean where they may drift. Ridley (1930) reported that these seeds were the most common disseminules on the beaches of Malaysia. While the intact seed coat provides excellent buoyancy and may drift for great distances, the embryo in these drift seeds is usually absent or nearly so. Long distance dispersal by ocean currents is doubtful. Fresh seeds, when freed from the decaying fruit, are covered with a white fleshy aril which is soon eroded. The seeds are quite oily and poisonous when fresh. They may be eaten only after they have been cooked in several changes of fresh water. The onion-flavored fruit is edible.

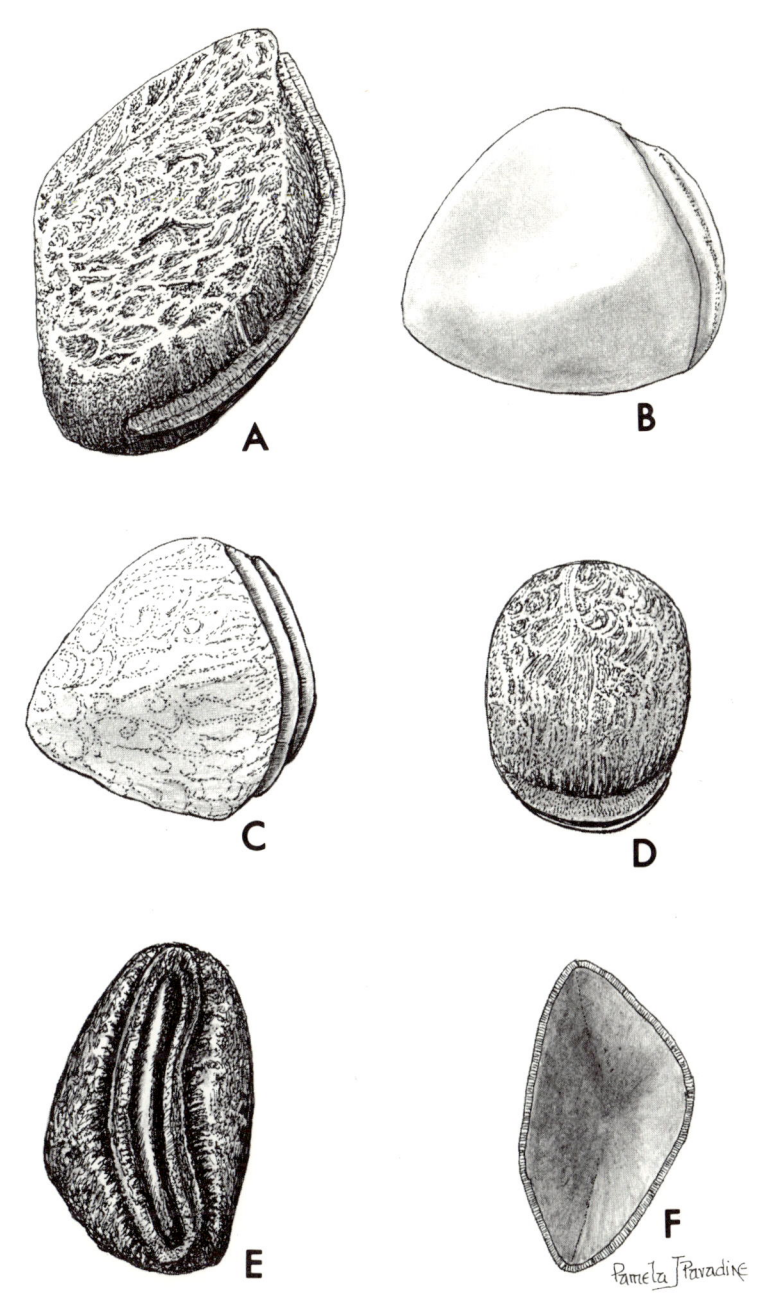

Flacourtiaceae

PANGIUM EDULE REINW.

Figure 39. *Pangium edule* seeds. A-D, lateral views; E, hilar view; F, cross section (X1). Seeds, Degener and Degener, beaches of Canton Island.

Guttiferae

CALOPHYLLUM CALABA L.
CALOPHYLLUM INOPHYLLUM L.

Disseminule: Mesocarp, rarely fruit.

Description: Mesocarp (B-F, H-J) 2 to 4 cm in diameter, globose, round in cross section, light to dark grayish ochre, surface smooth to slightly roughened, apex minutely beaked. Fruit (A, G) like mesocarp except that surface is wrinkled, because of the presence of the paper-thin exocarp.

Buoyancy factor: Corky mesocarp and space around seed.

Buoyancy: *C. calaba*—more than 1.5 years; *C. inophyllum*—not tested. The literature indicates that they are long range drifters.

Viability: *C. calaba*—about 50 percent of the seeds are viable; *C. inophyllum*—not tested. The literature indicates that they are viable.

Currents: *C. calaba*—tropical currents arising in the New World; *C. inophyllum*—tropical currents arising in the southeastern Asia and Pacific regions, perhaps the New World.

We are unable to separate the mesocarps of these two species. Some authors suggest that *C. calaba* mesocarps are smaller than *C. inophyllum* mesocarps. While the Old World mesocarps appear to be *C. inophyllum* the size variation of the New World mesocarps may indicate that other species are present. The three likely species are *C. calaba* (a native species), *C. inophyllum* (now a pantropic species), and *C. brasiliensis* Camb. (a native of Brazil that is now widely planted in the American tropics). *Calophyllum* trees are known as laurelwood or beauty-leaf, and they are planted for their attractive leaves and shade. *Calophyllum inophyllum* has the added advantage of being wind and salt spray resistant.

We do not agree with Guppy (1917) that mesocarps of *C. inophyllum* are more seaworthy than those of *C. calaba*. We believe that their range difference is due to the ability of *C. inophyllum* trees to live along the coast, while *C. calaba* trees live inland along rivers. Therefore, there is a much greater opportunity for *C. inophyllum* mesocarps to become stranded in a suitable habitat than for mesocarps of *C. calaba*. Like *Barringtonia asiatica*, *C. inophyllum* is one of the first trees to colonize a newly formed island in the Pacific region.

Guttiferae

CALOPHYLLUM SPP.

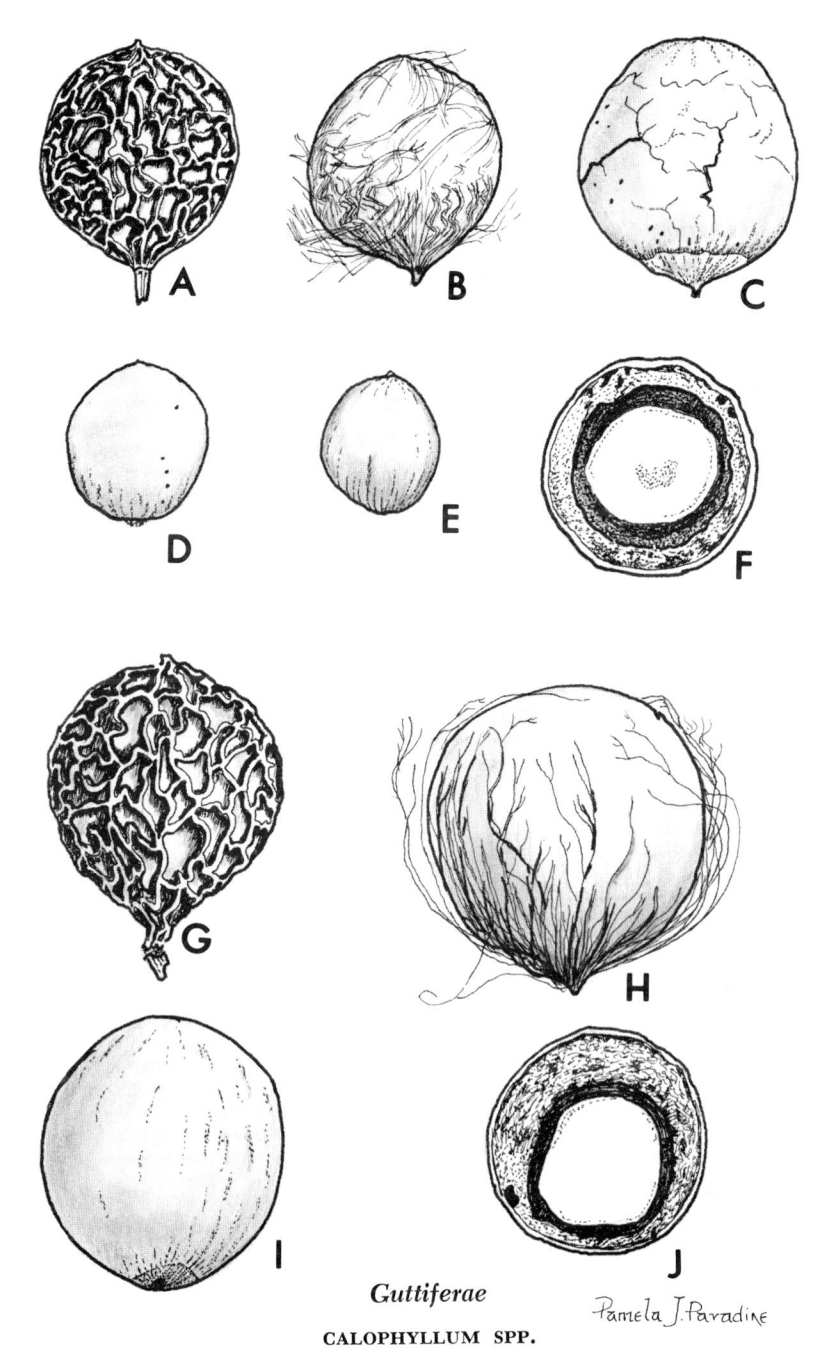

Figure 40. *Calophyllum calaba* (A-F); *C. inophyllum* (G-J). A, G, fruits; B-F, H-J, mesocarps. A-E, G-I, lateral views; F, J, cross section (X1). *C. calaba*, Gunn and Dennis, southeastern coast of Florida; *C. inophyllum* fruit, from tree; mesocarps, Degener and Degener, beaches of Canton Island.

(111)

Guttiferae

MAMMEA AMERICANA L., MAMMEE-APPLE

Disseminule: Mesocarp.
Description: Mesocarp (A-D) 5 to 7 cm long, 3 to 5 cm in diameter, ellipsoidal to oblong, rounded to compressed in cross section, tan to dark brown, surface fibrous. Fibrous surface may be concealed under remnants of the blackish exocarp.
Buoyancy factor: Empty space around seed.
Buoyancy: About 5 months.
Viability: None viable.
Currents: Tropical currents arising in the New World and perhaps in the Old World.

Mammee-apple, a native tree of the New World tropics, has been spread by man to the Old World tropics. Ocean currents could not have caused this spread, because the seeds are consistently dead in stranded endocarps. Also there is no transport current available to move disseminules from the New World to the Old World. Fresh fruits and fresh mesocarps are seldom buoyant. As the exocarp erodes and the seed shrinks, the mesocarp becomes buoyant. Fresh fruits are edible, although the flesh next to the mesocarp may be bitter and somewhat poisonous. The seeds are poisonous.

Guttiferae

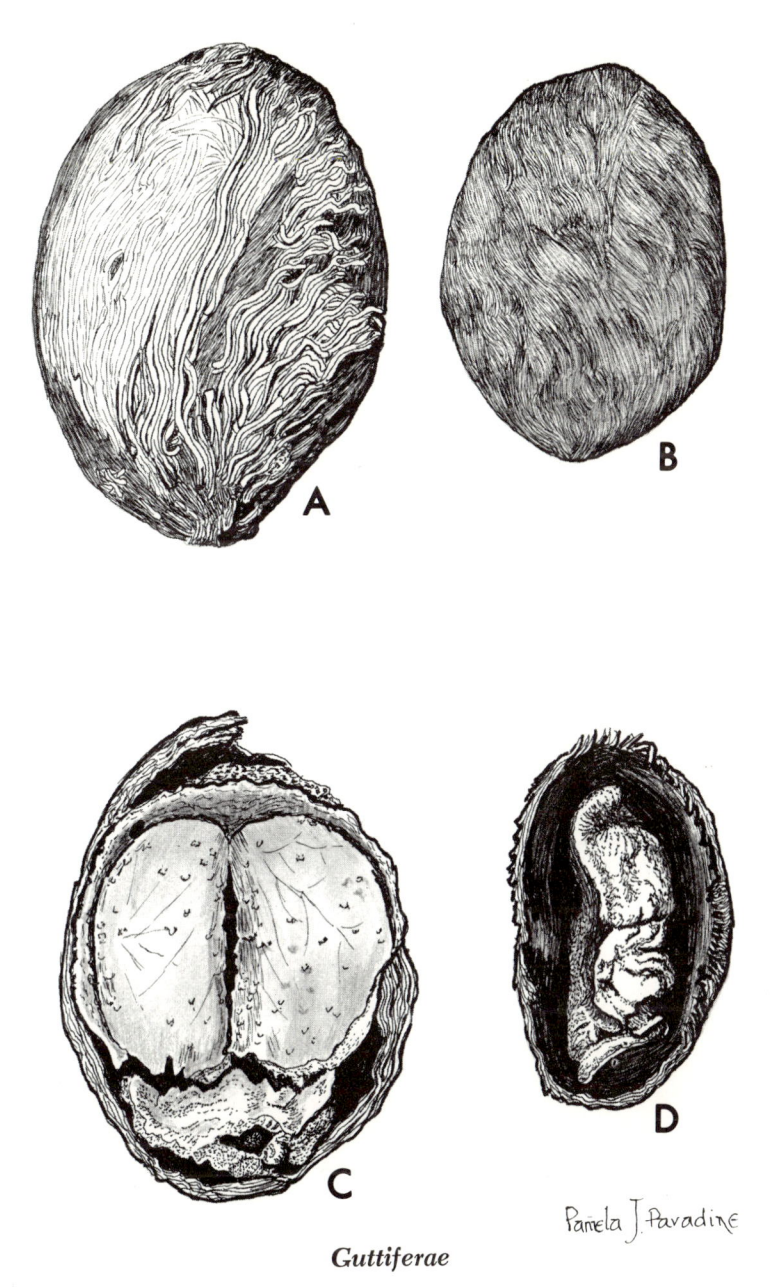

MAMMEA AMERICANA L.

Figure 41. *Mammea americana* mesocarps. A-B, lateral views; C, damaged mesocarp; D, cross section (X1). Mesocarps, Gunn and Dennis, southeastern coast of Florida.

Hernandiaceae

HERNANDIA NYMPHIIFOLIA (PRESL) KUBITZKI
HERNANDIA SONORA L.

Synonyms: *H. nymphiifolia* is *H. peltata* Meissner, and *H. sonora* is *H. guianensis* Aublet in the drift literature.

Disseminule: Endocarp or entire fruit (calyx absent from our disseminules).

Description: Endocarps of *H. nymphiifolia* (A-F) and *H. sonora* (G-L) are identical. Endocarp 1 to 2 cm in diameter, globose, round in cross section, dull tan to dull or lustrous reddish brown to black, surface smooth and bearing a faint encircling line. Remnants of the endocarp are tan and the amount varies from isolated patches to complete (G) with 6 to 8 corky ribs.

Buoyancy factor: Buoyant cotyledonary tissue.

Buoyancy: *H. nymphiifolia*—not tested; *H. sonora*—at least 2 years.

Viability: *H. nymphiifolia*—not tested; *H. sonora*—about 25 percent viable.

Currents: *H. nymphiifolia*—tropical currents arising between East Africa and Fiji; *H. sonora*—tropical currents arising in the New and Old World.

Our endocarps were examined by K. Kubitzki, Institut für Systematische Botanik, Universität München, who positively identified endocarps of *H. nymphiifolia* and observed that the American endocarps could be either *H. sonora* of the Antilles or *H. guianensis* of the American continent. We are following Standley and Steyermark (1946) who regarded these two species as one under the older name, *H. sonora*. This species is pantropic. Fresh fruits are almost encased by an inflated, globose calyx which gives the genus a distinctive character and ornamental value. According to Ridley (1930), fruits surrounded by the calyx float. The calyx remains upright in seawater, keeping the water out of the fruit cavity. Schimper (1891) credited fruit buoyancy to a buoyant layer between the endocarp and the seed. This statement is in conflict with the generic characters of the fruit.

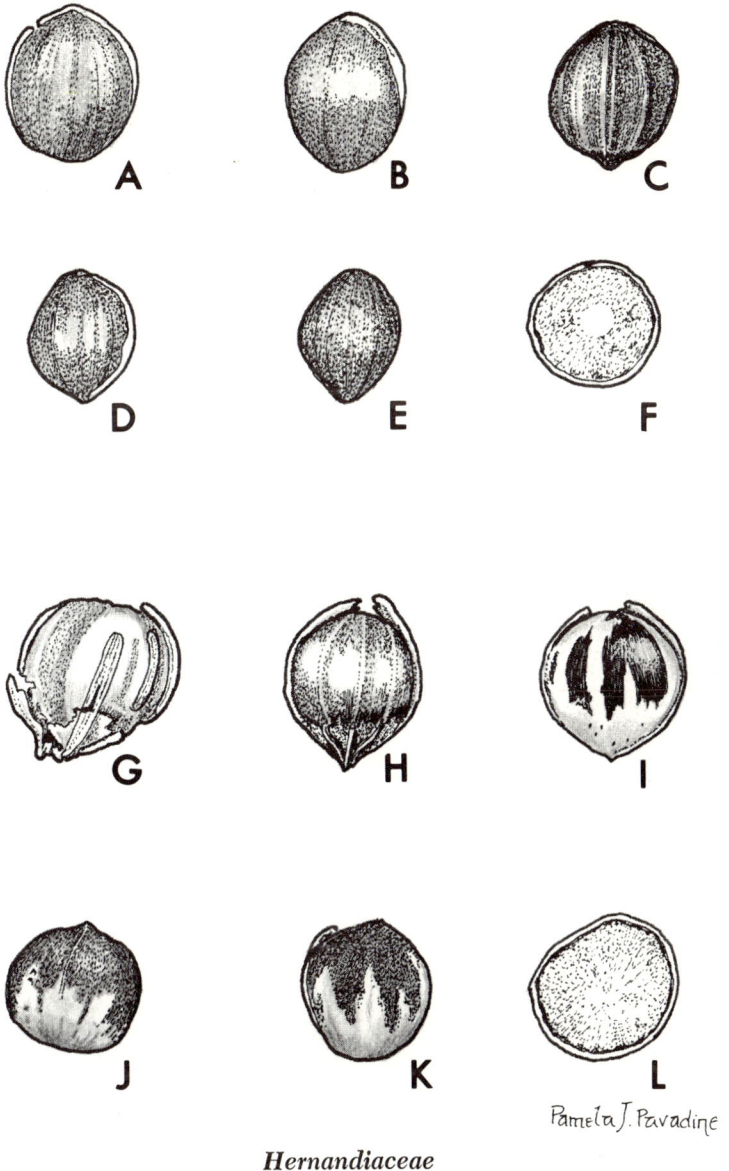

Hernandiaceae

HERNANDIA SPP.

Figure 42. *Hernandia nymphiifolia* (A-F); *H. sonora* (G-L). A-F, J-L, endocarps; G-I, fruits, A-E, G-K, lateral views; F, L, cross sections (X1). A-F, Degener and Degener, beaches of Canton Island; G-L, Gunn and Dennis, southeastern coast of Florida.

Humiriaceae

SACOGLOTTIS AMAZONICA MARTIUS

Disseminule: Endocarp, rarely fruit.
Description: Endocarp (B-H) 2 to 6 cm long, 2 to 4 cm in diameter, oblong, round in cross section, light to dark brown or grayish brown, surface slightly lumpy over cysts (cavities shown in G and H), if cysts are ruptured then surface irregularly pitted (F). Fruit (A) dark brown to blackish brown, often with a cracked surface.
Buoyancy factor: Empty resin cysts.
Buoyancy: At least 2 years.
Viability: About 30 percent of the fruits contain a viable seed.
Currents: Tropical currents arising in the New World.

These unusual endocarps, without a widely accepted English name[*], are produced by small trees which are native to the Amazon and Orinoco River basins and adjacent islands. Like *Merremia discoidesperma* drift seeds, these drift endocarps have a long history, because they drift to northern Europe. The first stranded endocarp illustration was published by Clusius (1605). Other early illustrations include those of Jonston (1622) and Petiver (1764). Yet this tree has never become established on Jamaica (Adams et al., 1972) or other Caribbean islands. No one has satisfactorily explained why this species has not colonized these islands. Sloane (1696), a Scottish medical doctor and botanist, studied the flora of Jamaica for 15 years and associated the stranded endocarps from Jamaican beaches with those found on beaches of northern Scotland and adjacent islands. However, he was unable to identify the endocarps and was unable to explain how they arrived on these widely separated beaches. Morris (1889, 1895) was the first to identify these drift endocarps, about 300 years after the Clusius publication. He did this by comparing an illustration in Urban (1877) with one made by Cruger.

Endocarps of the related genus *Vantanea* float, but apparently they are not found too far beyond their northern South American homeland. *Vantanea guianensis* Aublet (I) with its conspicuously rugose endocarp represents this genus. The family was recently monographed by Cuatracasas (1961).

[*] The Spanish common name, *cojon de burro,* while apt is not a name we wish to perpetuate in English.

Humiriaceae

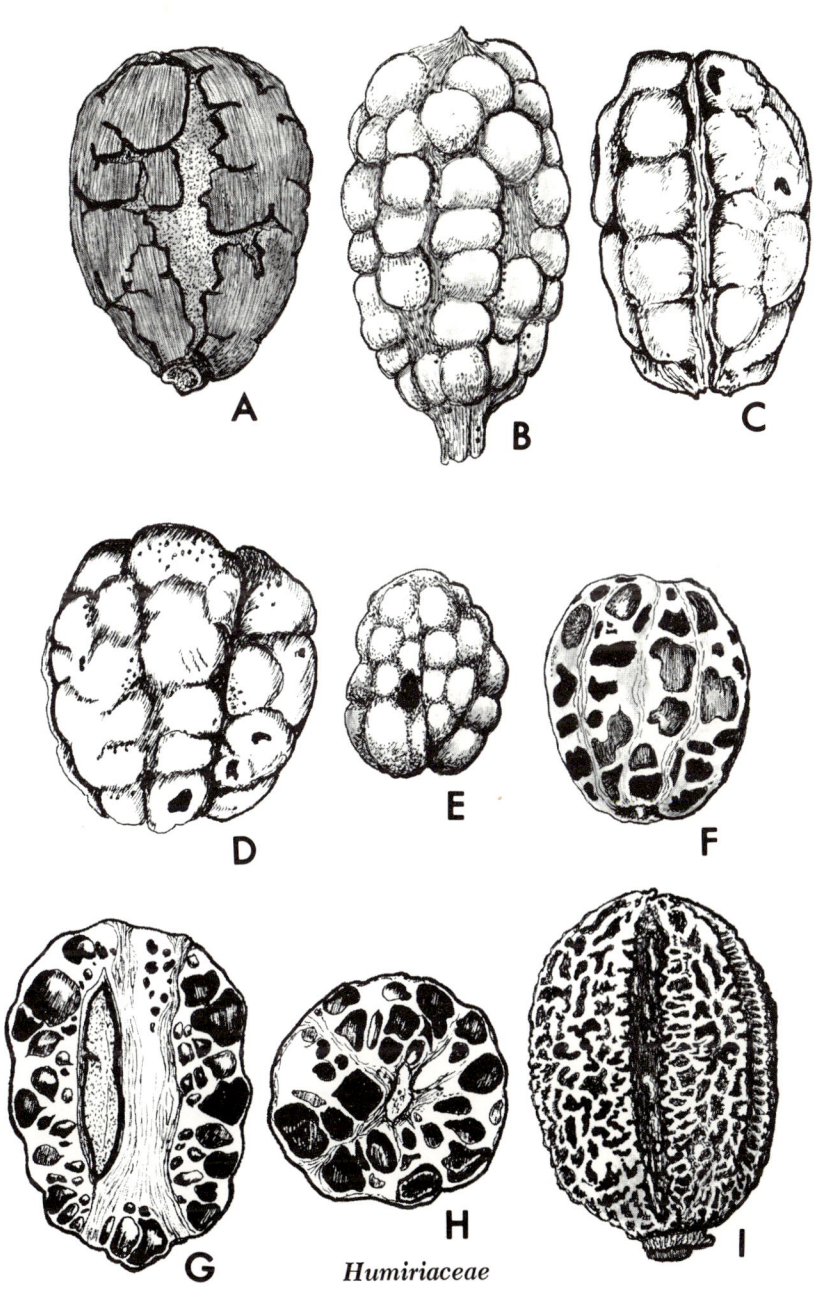

SACOGLOTTIS AMAZONICA MARTIUS
VANTANEA GUIANENSIS AUBLET

Figure 43. *Sacoglottis amazonica* (A-H); *Vantanea guianensis* (I). A, fruit; B-I, endocarps. A-F, I, lateral views; G, longitudinal view; H, cross section (X1). A-H, Gunn and Dennis, southeastern coast of Florida; I, from tree.

Icacinaceae

CALATOLA COSTARICENSIS STANDLEY

Disseminule: Endocarp.
Description: Endocarp (A-D) 4 to 7 cm long, 3 to 5 cm in diameter, fusiform, round in cross section, tan, surface prominently sculptured, bearing an encircling ridge (C) and narrow prominent longitudinal ridges with minor cross ridges in a variable pattern.
Buoyancy factor: Empty endocarp.
Buoyancy: At least 1.5 years.
Viability: Seed absent.
Currents: Tropical currents arising in the New World.

This tree is a little known species of limited distribution in Central America and adjacent northern South America. The sculptured endocarps, weathered free of the pulpy exocarp before or during drifting, are consistently empty or nearly so. While the endocarps may drift for more than one year, there is little chance of spreading by drifting. The species has not been spread by man to any extent, because it has no economic value. Endocarps are brittle, only 0.5 mm thick, and are often stranded in a damaged condition (C). Johnston (1949) found endocarps of *C. costaricensis* and one endocarp of *C. venezuelana* Pittier on the beaches of San Jose Island, situated on the Pacific side of Panama.

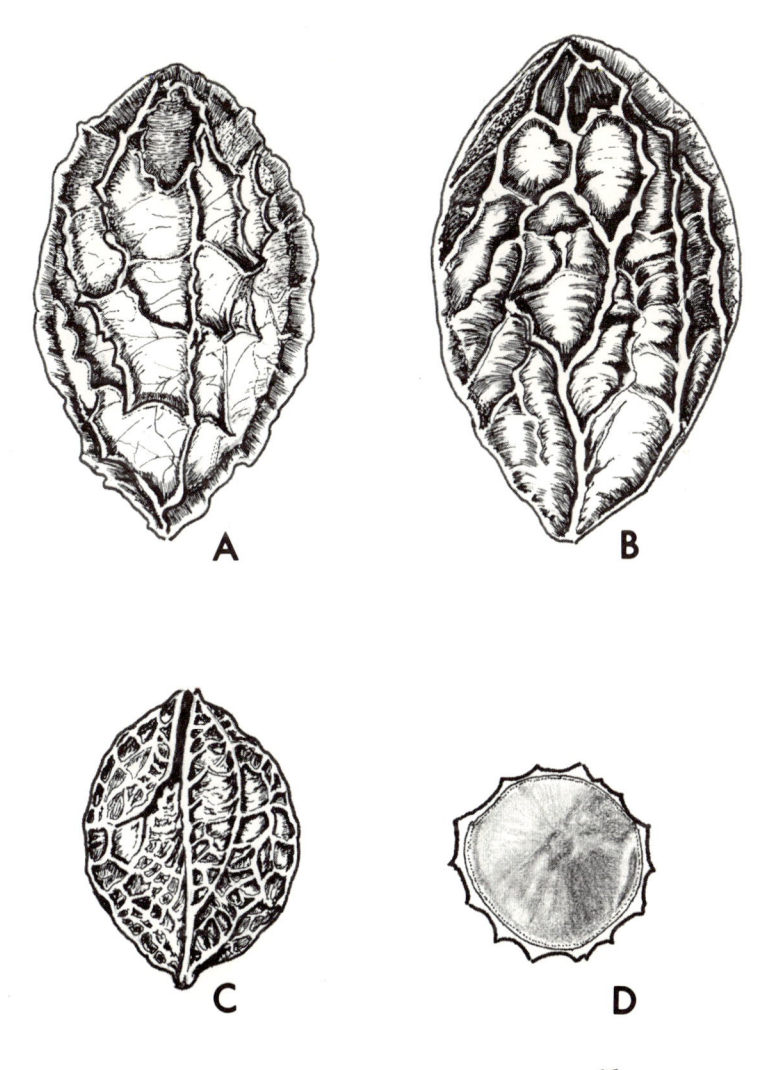

Icacinaceae

CALATOLA COSTARICENSIS STANDLEY

Figure 44. *Calatola costaricensis* endocarps. A-C, lateral views; D, cross section (X1). Endocarps, Dennis, southeastern coast of Florida.

Juglandaceae

CARYA AQUATICA (MICHX. F.) NUTT., WATER HICKORY
CARYA GLABRA (MILL.) SWEET, PIGNUT
CARYA ILLINOENSIS (WANG.) K. KOCH, PECAN
CARYA TOMENTOSA NUTT., MOCKERNUT

Disseminule: Endocarp, rarely fruit.
Description: See species discussions.
Buoyancy factor: Empty chambers or partially filled seed cavities.
Buoyancy: *C. aquatica*—more than 8 months; other species—not tested.
Viability: Few viable.
Currents: Gulf Stream and Gulf of Mexico currents.

Carya spp. are native United States trees whose endocarps are primarily brought to the Gulf of Mexico and the Atlantic Ocean by south and east flowing rivers, especially the Mississippi River. The endocarps are stranded with tropical disseminules on southern and eastern United States beaches. Except for *C. illinoensis,* the endocarps are quite durable, up to 5 mm thick. The endocarp of *C. illinoensis* is about 1 mm thick. Because the endocarps of *C. glabra* and *C. tomentosa* usually have an open suture, the seeds are dead. Sutures of *C. aquatica* and *C. illinoensis* remain closed, thus their seeds may be stranded alive. Our collections were identified by Wayne E. Manning, Bucknell University.

Carya aquatica endocarp (B-D) 2 to 3.5 cm long, 1.5 to 3 cm wide, subglobose to obovoid, compressed in cross section, light to dark brown, surface wrinkled with prominent ridge-like suture encircling length, apex short beaked. Entire fruit (A) bearing a smooth, black, 4-parted exocarp. These endocarps are more common than all of the others combined.

Carya glabra endocarp (E-G) 2 to 3 cm long, 2 to 2.5 cm wide, obovoid, slightly compressed in cross section, dark tan, surface minutely rugose, apex beaked (beak may be eroded). Entire fruit (partially shown in E) bearing a smooth black, 4-parted exocarp.

Carya illinoensis endocarp (I-K) 2.5 to 4 cm long, 1.5 to 3.5 cm in diameter, ellipsoidal, round in cross section, light to dark brown, surface smooth, apex and often base short acuminate. Entire fruit (H) bearing a smooth black, 4-parted exocarp.

Carya tomentosa endocarp (L-N) 3.5 to 5 cm long, 1.5 to 2.5 cm wide, globose to ellipsoidal, slightly compressed in cross section, grayish brown, surface smooth or nearly so, rounded at apex.

Juglandaceae

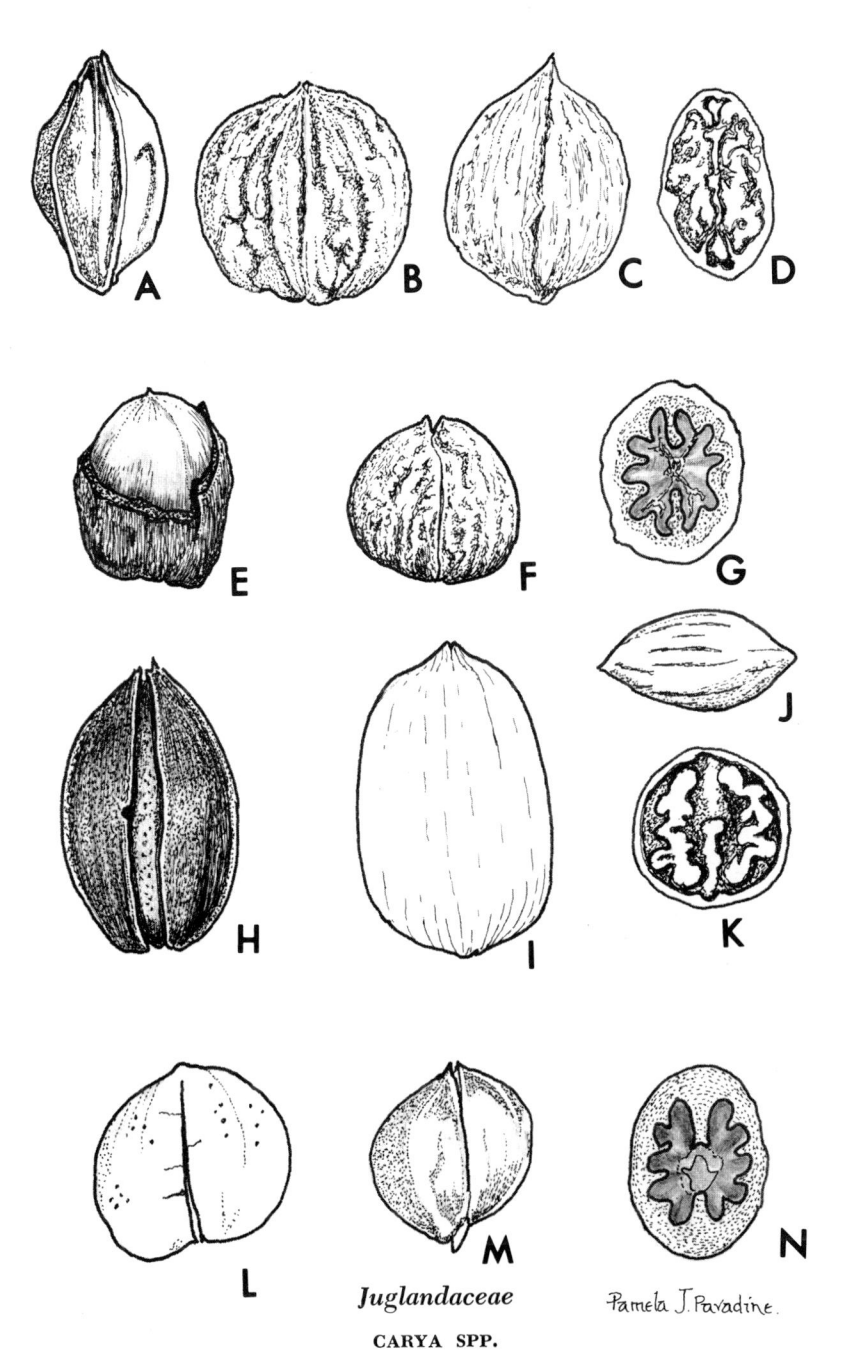

CARYA SPP.

Figure 45. *Carya aquatica* (A-D); *C. glabra* (E-G); *C. illinoensis* (H-K); *C. tomentosa* (L-N). A, E, H, fruits or partial fruits; B-D, F-G, I-N, endocarps. A-C, E-F, H-J, L-M, lateral views; D, G, K, N, cross sections (X1). Fruits and endocarps, Gunn and Dennis, southeast coast of Florida and United States coast along the Gulf of Mexico.

(121)

Juglandaceae

JUGLANS CINEREA L., WHITE WALNUT

JUGLANS JAMAICENSIS C. DC.

JUGLANS NIGRA L., BLACK WALNUT

JUGLANS REGIA L., ENGLISH WALNUT

Disseminule: Endocarp. Our collections were identified by Wayne E. Manning, Bucknell University.
Description: See species discussions.
Buoyancy factor: Empty chambers or partially filled seed cavities.
Buoyancy: See species discussions.
Viability: No viable seeds found. Endocarps composed of 2 halves which are often partially open, allowing seawater to enter and kill the seed.
Currents: Gulf Stream and Gulf of Mexico currents, perhaps other currents.

Juglans cinerea endocarp (A-B) 3.5 to 5 cm long, 2 to 2.5 cm wide, ellipsoidal, round in cross section, light to dark brown, surface deeply furrowed, apex pointed, base somewhat pointed. White walnut is a native tree of temperate United States. Endocarps are transported by rivers to the Gulf of Mexico and the Atlantic Ocean and found on United States beaches that regularly receive tropical disseminules. Buoyancy—about 3 months.

Juglans jamaicensis endocarp (C-D) 2 to 5 cm long, 2 to 4 cm broad at base, ovate, round in cross section, light to dark brown, surface furrowed, apex often acuminate, base concave. This tree is a native of the Caribbean region. Its endocarps float for at least 22 months, and they drift in tropical currents arising in the Caribbean region. It is possible that some of these endocarps are from *J. insularis* Griseb. These endocarps are reported to drift (Adams et al, 1972).

Juglans nigra endocarp (E-F) 3 to 4 cm in diameter, subglobose, slightly compressed in cross section, blackish to light brown, surface furrowed, short tapered to apex, rounded at base. Black walnut is a native tree of temperate United States, and its endocarps are transported by rivers to the Gulf of Mexico and the Atlantic Ocean. The endocarps are found on United States beaches which regularly receive tropical disseminules. Buoyancy—4 to 5 months.

Juglans regia endocarp (G-H) 3 to 5 cm long, 2.5 to 3 cm wide, oblong, round in cross section, brown, surface bearing shallow irregular grooves, apex short minute-pointed, base flat or tapered. These endocarps are more fragile (thinner walled) than the others. The English walnut is a widely planted tree, and its endocarps float for about 15 months. They may be found in both New and Old World currents.

Juglandaceae

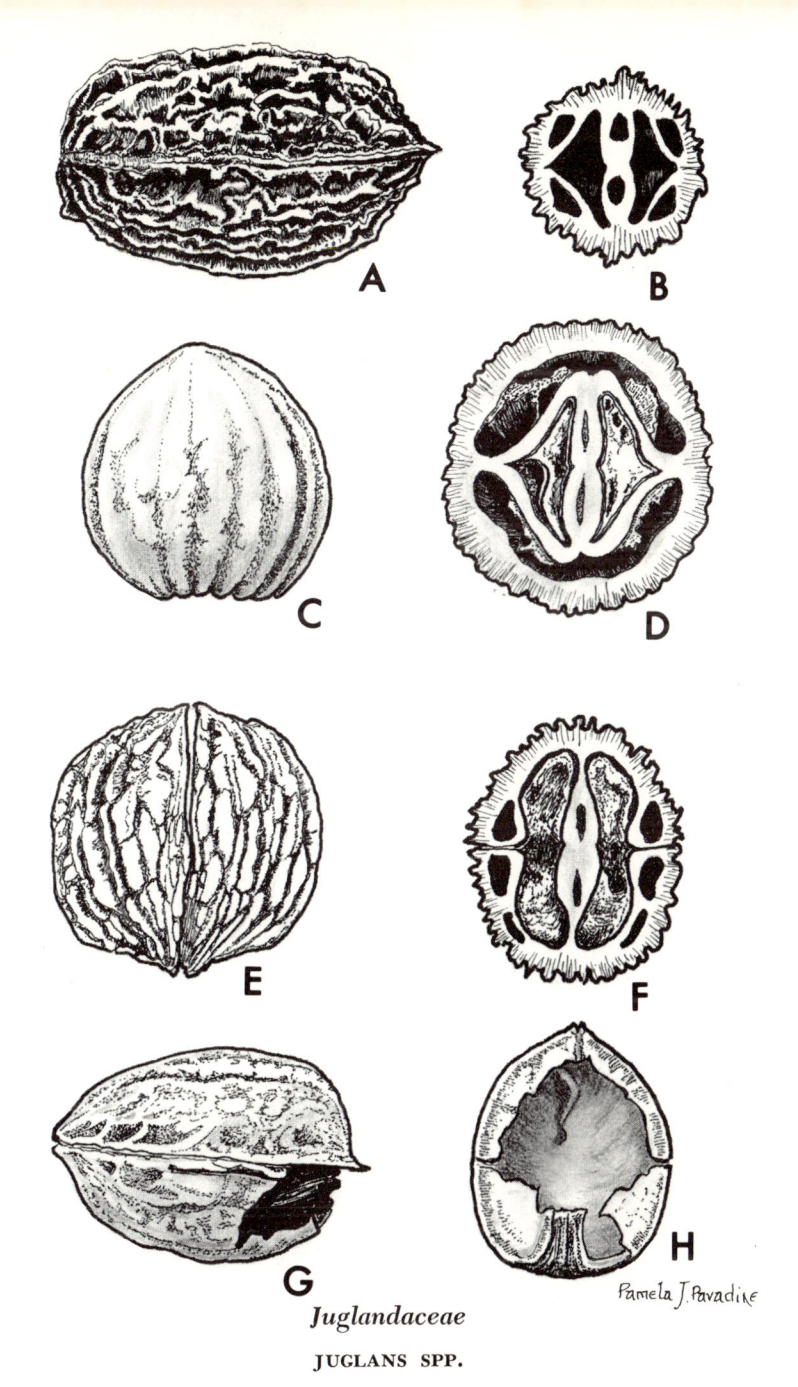

JUGLANS SPP.

Figure 46. *Juglans cinerea* (A-B); *J. jamaicensis* (C-D); *J. nigra* (E-F); *J. regia* (G-H). A-H, endocarps. A, C, E, G, lateral views; B, D, F, cross sections; H, longitudinal section (X1). Endocarps, Gunn and Dennis, southeastern coast of Florida and the United States Gulf coast.

(123)

Lecythidaceae

GRIAS CAULIFLORA L., ANCHOVY-PEAR

Disseminule: Mesocarp, rarely entire fruit.
Description: Mesocarp (A-E) 7 to 9 cm long, 2 to 4 cm in diameter, ellipsoidal, rounded in cross section, surface ribbed, ribs tan, interstices chocolate brown to tan, bearing 8 prominent parallel ribs, cross ribs present on deeply eroded specimens.
Buoyancy factor: Corky mesocarp and perhaps empty space around seed.
Buoyancy: At least 19 months.
Viability: About 10 percent viable.
Currents: Tropical currents arising in the New World.

The anchovy-pear, a native tropical American tree, is usually found along rivers or in marsh forests where it may form large colonies. Its mesocarps float in fresh water, though naked seeds are not buoyant. The species is spread by rivers. Most of these drifting mesocarps bear germinating seeds, or seeds which are dead or dying. Those intact mesocarps which make it to the ocean have little chance to be carried to a suitable habitat for colonization. Like the *Barringtonia asiatica* embryo, the embryo of the anchovy-pear has undifferentiated cotyledons. This is an unusual condition among seed-bearing plants.

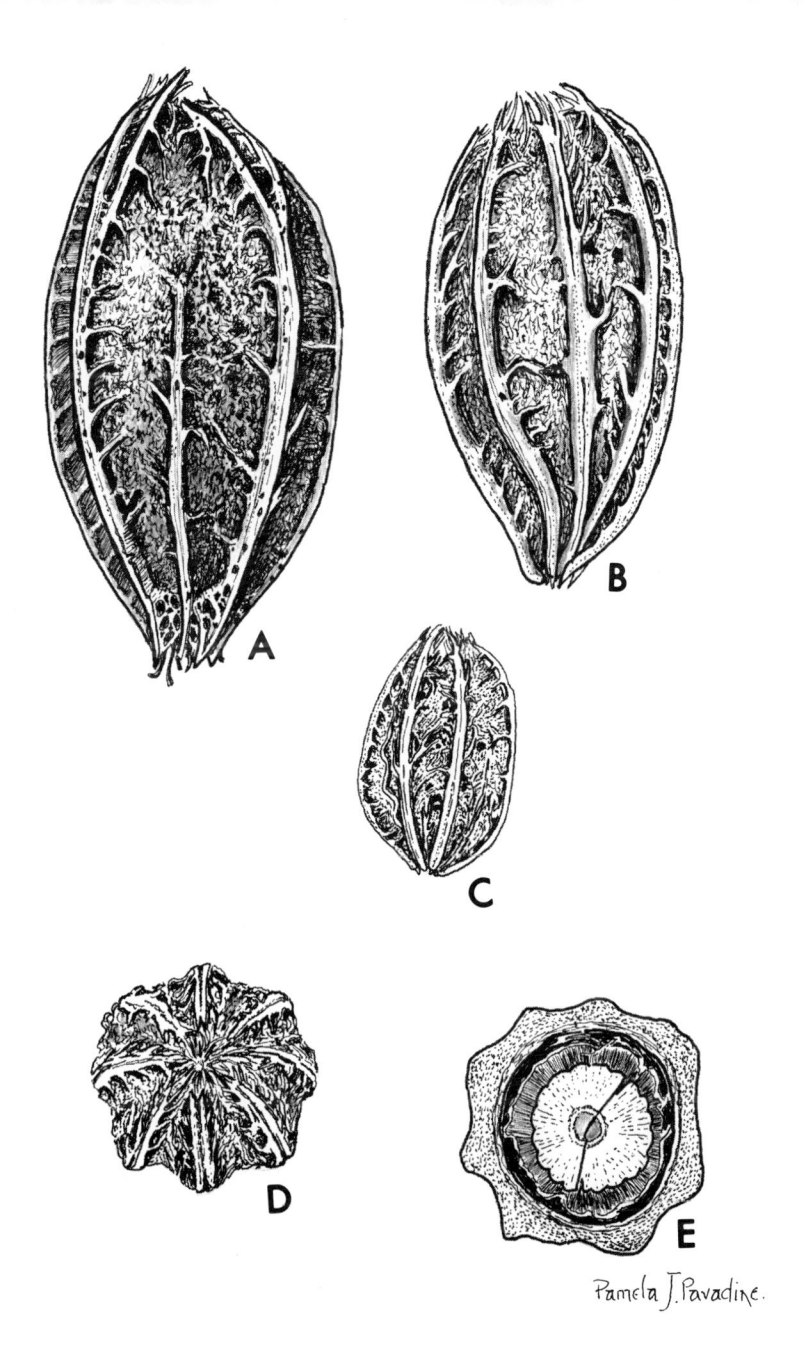

Lecythidaceae

GRIAS CAULIFLORA L.

Figure 47. *Grias cauliflora* mesocarps. A-C, lateral views; D, apical view; E, cross section (X1). Mesocarps, Gunn and Dennis, southeastern coast of Florida.

Leguminosae

ANDIRA GALEOTTIANA STANDLEY
ANDIRA INERMIS (W. WRIGHT) H.B.K., CABBAGEBARK

Disseminule: *A. galeottiana*—fruit; *A. inermis*—mesocarp, rarely fruit.
Description: *A. galeottiana* fruit (A-B) 6 to 8 cm long, 5 to 6 cm in diameter, elongate, slightly compressed in cross section, brown to tan, surface nearly smooth and bearing a poorly defined indented suture which encircles the length of the fruit. *A. inermis* mesocarp (D-H) 3 to 6 cm long, 2 to 4 cm in diameter, globose, ellipsoidal, or elongate, round in cross section, tan to brown, surface fibrous bearing a prominent (occasionally eroded) ridge-like suture encircling length of mesocarp, faint irregular ridges, and occasionally fine pits. Fruit (C) similar to mesocarp, except outer surface is smooth when fresh, eroding as fruit drifts.
Buoyancy factor: Fibrous mesocarp and space around seed.
Buoyancy: *A. galeottiana*—not tested; *A. inermis*—at least 2 years.
Viability: *A. galeottiana*—not tested; *A. inermis*—about 25 percent viable.
Currents: *A. galeottiana*—Gulf of Mexico; *A. inermis*—tropical currents arising in the New World and along the west coast of Africa.

Andira galeottiana is a little known small native tree of inland southern Mexico. The fruits are carried by rivers to the Gulf of Mexico where they may drift for a limited time. Fruits have been collected along the southern Mexican coast and along beaches of Brazos Island, Texas (Gunn and Dennis, 1973), but not from beaches of the Yucatan.

The common name for *A. inermis* is derived from the cabbage aroma emitted from cut bark. Cabbagebark is a native tree of tropical America, and it has been spread by man to western Africa. Drifting was not responsible for this spread, because there is no transport current. While cabbagebark disseminules are frequently found stranded on New World beaches, the seeds are often dead. Even though the fibrous mesocarp is thick and quite buoyant, the fibrous tissue is not hard. Thus the buoyant layer is easily eroded or pitted, permitting seawater to penetrate the seed cavity and kill the seed. The seeds are thought to be poisonous (Standley, 1924).

Leguminosae

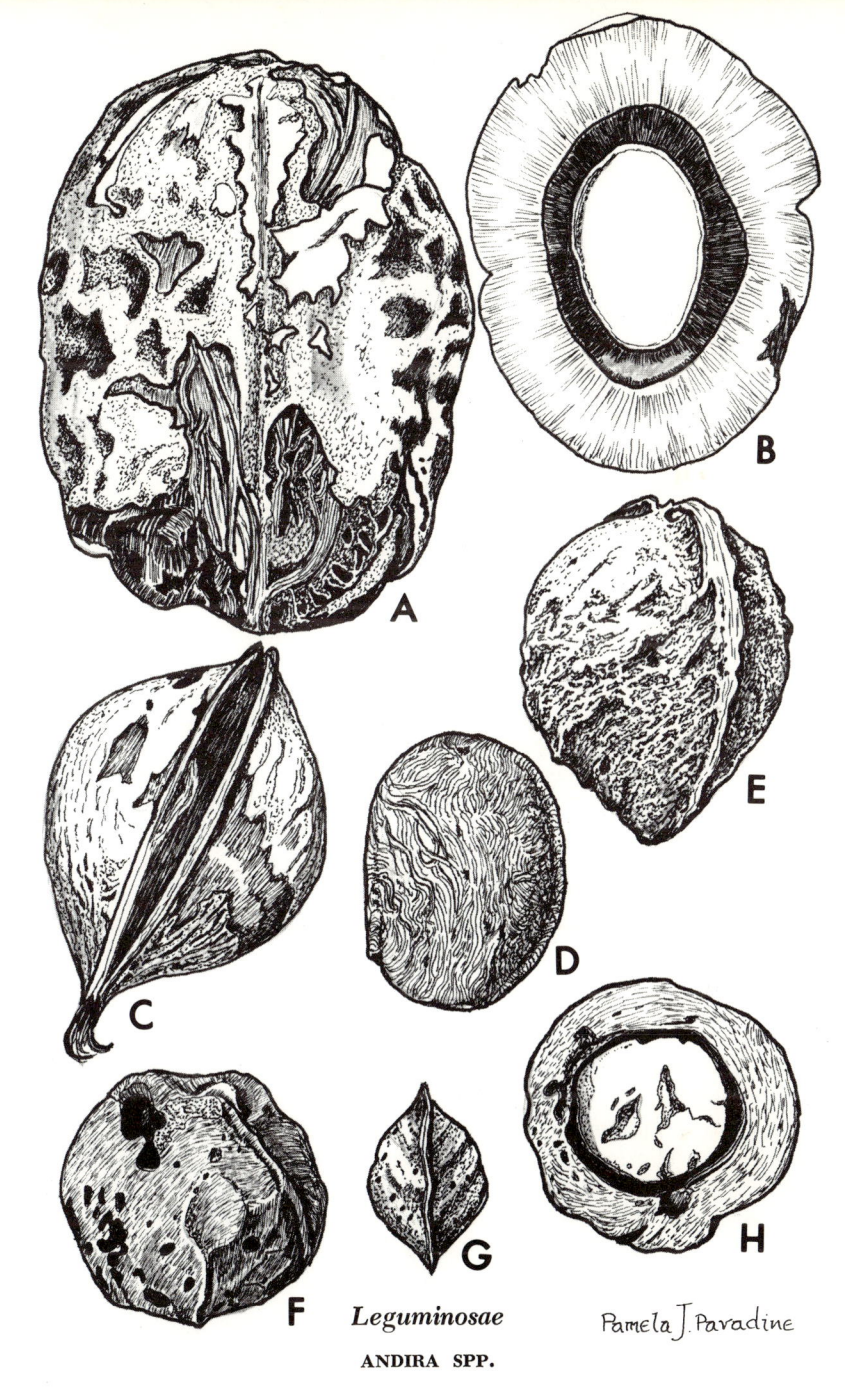

ANDIRA SPP.

Figure 48. *Andira galeottiana* (A-B); *A. inermis* (C-H). A-C, fruits; D-H, mesocarps. A, C-G, lateral views; B, H, cross sections (X1). *A. galeottiana*, Dennis, Brazos Island, Texas; *A. inermis*, Gunn and Dennis, southeastern coast of Florida.

(127)

Leguminosae

CAESALPINIA BONDUC (L.) ROXB., GRAY NICKERNUT
CAESALPINIA MAJOR (MEDIKUS DANDY & EXELL, YELLOW NICKERNUT

Synonyms: *C. crista* L., *C. bonducella* (L.) Fleming, and these species names in *Guilandina* are synonyms of *C. bonduc* in the drift literature.
Description: *C. bonduc* seed (B-G) 1.5 to 2.5 cm in diameter, subglobose, more or less round in cross section, lustrous to dull light olive to silver gray to grayish-yellow, surface smooth though bearing numerous faint concentric fracture lines (D-G). Hilum conspicuous brownish area (B-C). *C. major* seed (H) similar, except a bright yellow color.
Buoyancy factor: Intercotyledonary cavity.
Buoyancy: At least 2 years.
Viability: Most seeds viable.
Currents: Most tropical currents.

The gray nickernut is one of the nicer, long ranging tropical drift seeds. Its drift history may be traced to Clusius (1605), who reported seeds on northern European beaches. The species is a native of southeast Asia, but now has attained a pantropic distribution, primarily by drifting. The seeds are produced on spiny, trailing shrubs, which form dense thickets just beyond the high tide zone. Thus it is not surprising that this is a common drift seed. The bitter tasting embryo has been much used in various remedies, especially for fever control, hence the common name fevernut. The seeds are also used as marbles and in seed jewelry.

The yellow nickernut is not a littoral shrub, and its seeds are rarely found stranded. When found on New World beaches, the seeds are highly prized, as they should be.

The chocolate brown seeds (I-K) have not been identified. We believe that they are members of the genus *Caesalpinia,* or at least a closely related genus. These seeds are more frequently found along beaches of the Gulf of Mexico than they are along the southeastern coast of Florida.

Other seeds in the genus drift, especially species from southeast Asia. These seeds bear the characteristic faint concentric fracture lines.

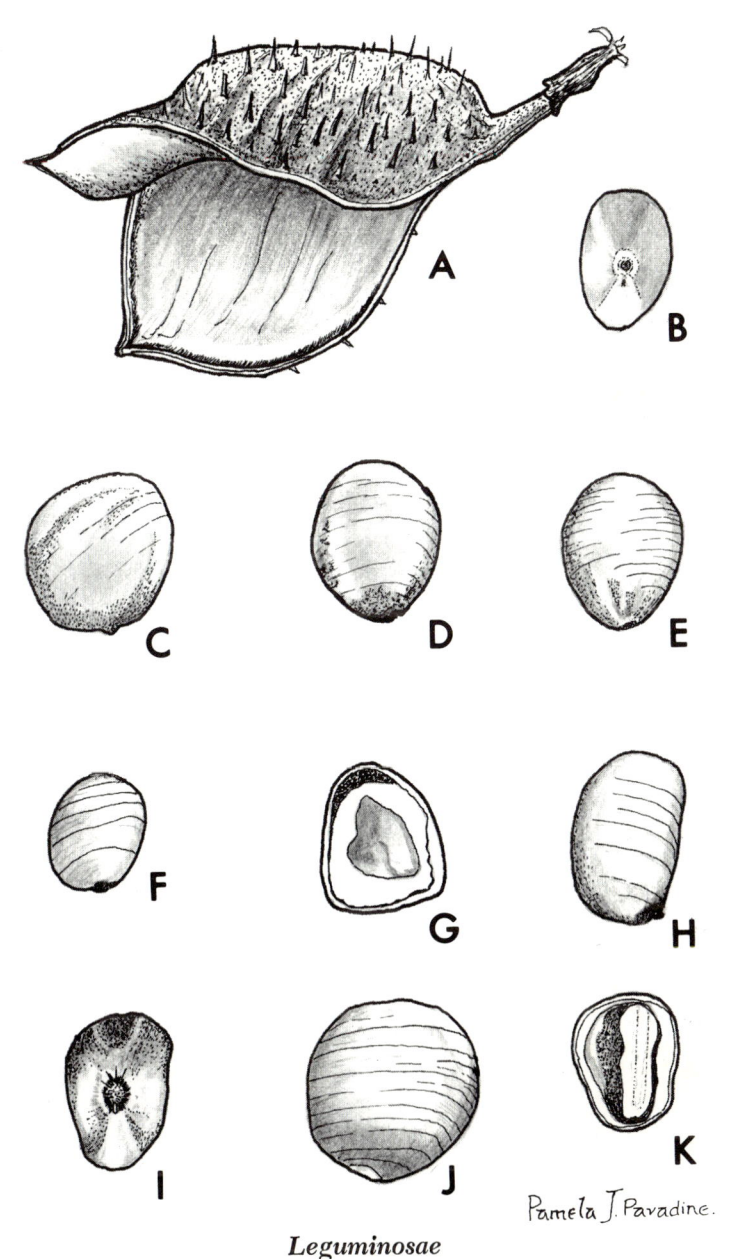

Leguminosae

CAESALPINIA SPP.

Figure 49. *Caesalpinia bonduc* (A-G); *C. major* (H); *C.* sp. ? (I-K). A, fruit; B-K, seeds. A, C-F, H, J, lateral views; B, I, hilar views; G, K, cross sections (X1). Fruit, from vine: seeds B-H Gunn and Dennis, southeastern coast of Florida; seeds I-K, Dennis, Gulf coast of Texas.

(129)

Leguminosae

CANAVALIA CATHARTICA THOUARS
CANAVALIA NITIDA (CAV.) PIPER
CANAVALIA ROSEA (SW.) DC., BAY-BEAN

Synonyms: See species discussions.
Disseminule: Seed.
Description: See species discussions.
Buoyancy: Impermeable to water for at least 1.5 years for the 3 named species.
Viability: *C. rosea*—most seeds viable; other species—not tested.
Currents: Seeds of *Canavalia* spp. are found in most tropical currents.

Sauer (1964) monographed the genus, and we are following his nomenclature. He is one of the few monographers to evaluate the impact drifting has had on distribution and phylogeny.

Canavalia cathartica seed (A) up to 20 mm long, about 12 mm wide, elliptical, nearly round in cross section, reddish brown, smooth, bearing a prominent hilum about 14 mm long. Sauer reported that the seeds were not buoyant but were impermeable to water for more than 1.5 years. Otto and Isa Degener collected drift seeds from beaches of Canton Island.

Canavalia nitida seed (D) about 25 mm long, about 15 mm wide, elliptical, compressed slightly in cross section, dark wine, red, black, or dark tan, smooth, bearing a black linear hilum about 25 mm long (longer than the length of the seed). This species is a native of the West Indies. Sauer reported the seeds to be impermeable to water for at least 1.5 years and not buoyant. We have collected drift seeds from the beaches of southern United States.

Canavalia rosea (formerly *C. maritima* (Aublet) Thouars and *C. obtusifolia* DC. of the drift literature) seed (C) 12 to 20 mm long, 10 to 13 mm wide, elliptical, slightly compressed in cross section, brown with darker mottles, smooth, bearing a brown hilum 7 to 8 mm long. Both Guppy (1917) and Sauer (1964) considered this species to be the prime progenitor of the other species in the genus. A ubiquitous strand vine, bay-bean forms vast tangled masses over the beaches of the tropics.

Other species produce drift seeds, including *C. bonariensis* Lindley of southern South America and *C. sericea* A. Gray of the Pacific Ocean region. An unidentified Canavalia stranded on the beaches of Florida is shown in E. Seeds are about 25 mm long, about 17 mm wide, elliptical, slightly compressed in cross section, brown, smooth, and bearing an oblong hilum about 18 mm long.

Leguminosae

CANAVALIA SPP.

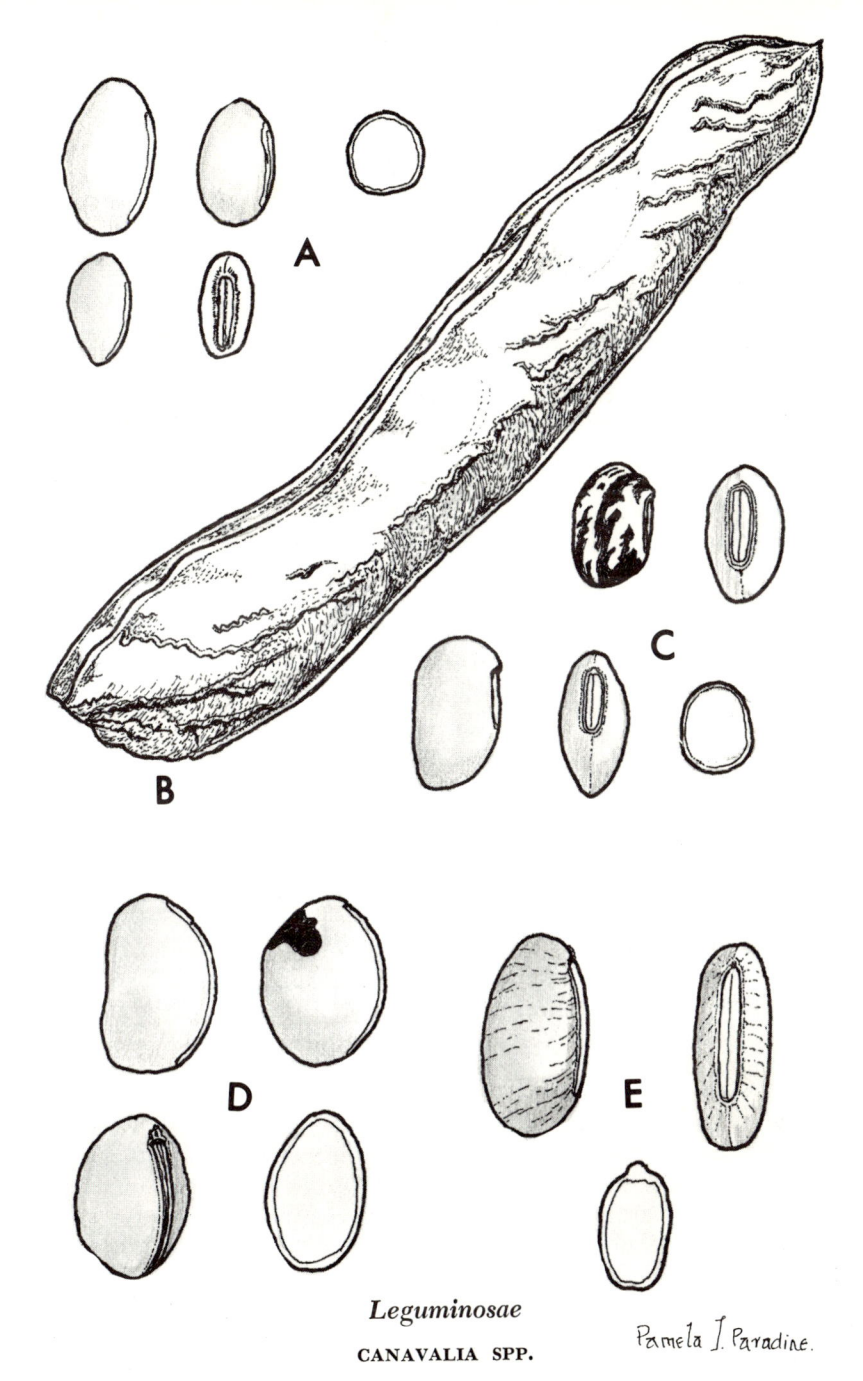

Figure 50. *Canavalia cathartica* (A); *C. rosea* (B-C); *C. nitida* (D); *C.* sp. (E). A, C-E, seeds in lateral, hilum, and cross section views; B, pod in lateral view (X1). *C. cathartica,* Degener and Degener, beaches of Canton Island; other species, Gunn and Dennis, southeastern coast of Florida.

Leguminosae

CASSIA FISTULA L., GOLDEN SHOWER

CASSIA GRANDIS L. F., PINK SHOWER

Disseminule: Fruit or fruit segment.

Description: *C. fistula* fruit (A-D) 30 to 60 cm long, about 2 cm in diameter, cylindrical, round in cross section, blackish brown, surface smooth or nearly so, short acuminate at apex, tapering to base, suture (B) an inconspicuous groove with flat margins, interior divided into many 1-seeded compartments (C). *C. grandis* fruit (F-H, J) 30 to 80 cm long, 2 to 3 cm in diameter, cylindrical, compressed in cross section, dark brown, surface moderately roughened, short acuminate at apex, tapering to base, suture (F) a conspicuous groove with thickened margins, interior divided into many 1-seeded compartments (G).

Buoyancy factor: Space within each seed chamber.

Buoyancy: *C. fistula*—at least 2 years; *C. grandis*—at least 6 months.

Viability: Most seeds viable for both species.

Currents: Tropical currents arising in the New World and perhaps Old World tropical currents.

Contrary to our buoyancy test results, *C. fistula* fruits are not long distance drifters. The Norwegian records of stranded *C. fistula* fruits are accurate, though contrary to the literature, they did not drift to Norway via the Gulf Stream (Guppy, 1917; Gunnerus, 1765; Lindman, 1882; Linnaeus, 1789; Sernander, 1901; and Ström, 1792). Like the cashew (*Anacardium occidentale* L.) fruits, *C. fistula* fruits were beached as the result of shipping accidents. Our evidence is 1) neither fruit has been reported from any beach served by the Gulf Stream between Norway and Florida; 2) both fruits are infrequent to rare on Florida beaches; and 3) both fruits were items of commerce during the seventeenth and eighteenth centuries. Brazilian *C. fistula* fruits were used in preparing a laxative. Hemsley (1885) reported that a fruit containing viable seeds was collected by Martins on a beach at Montpellier, France. He used this as evidence that fruit floated from the Caribbean region through the Straits of Gibraltar to the Montpellier beach. It is much simpler to credit the source of the fruit to North Africa, or to believe Martins' (1857) explanation that the fruit drifted from Marseilles. Furthermore, *C. fistula*, a native of the Indian region, was not reported in the drift of this region (Fairchild, 1943, Guppy, 1890, 1906, 1917; Hemsley, 1885; and Schimper, 1891). Muir (1937) reported finding pieces of the fruit on the Riversdale coast, and he noted that the plant was cultivated in Natal and Madagascar.

Fruits of *C. grandis*, a native New World tropics, are rarely found on southeastern Florida beaches. We believe that the fruits of *C. grandis* are usually too heavy to drift.

Leguminosae

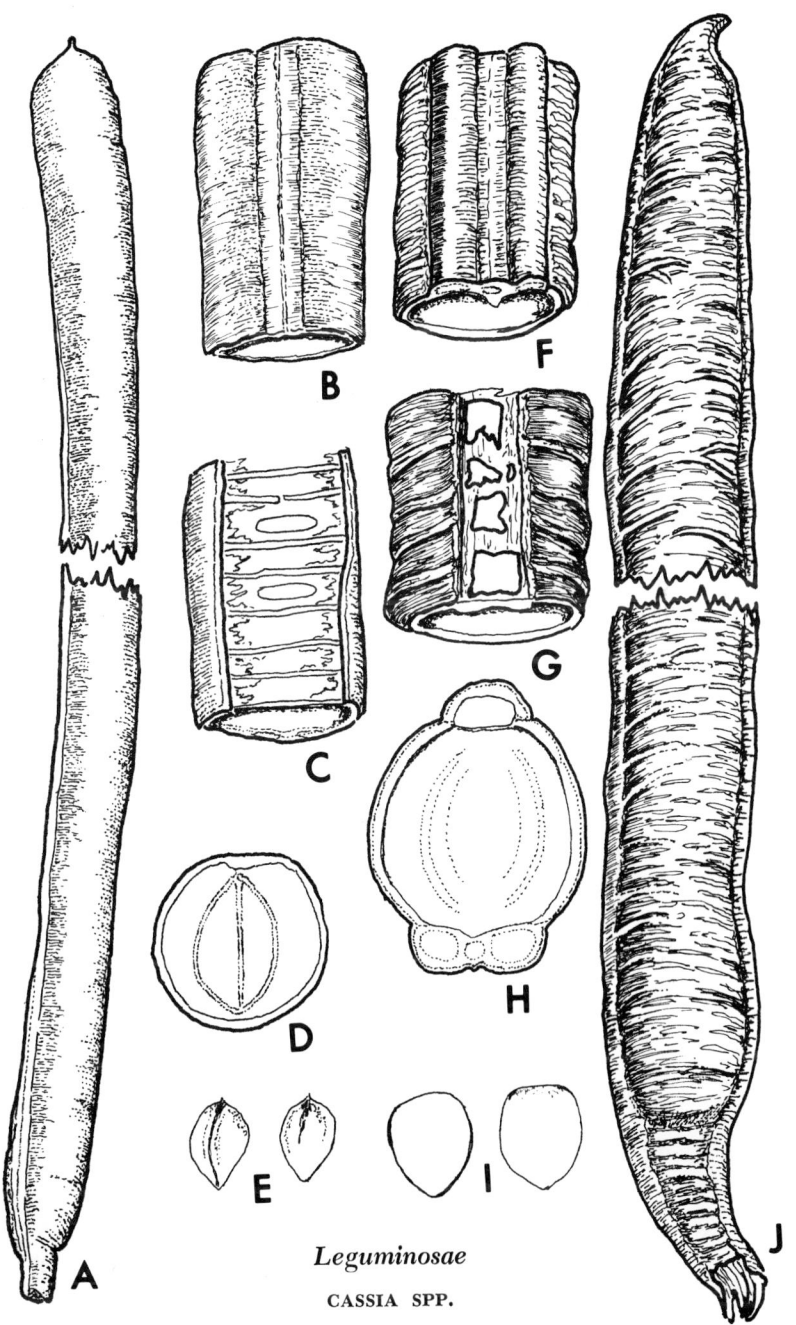

CASSIA SPP.

Figure 51. *Cassia fistula* (A-E); *C. grandis* (F-J). A, J, fruits (broken area indicates that fruits even though drawn at X½ are too long to fit the plate); B-C, F-G, fruit sections showing sutures and separate seed chambers; D, H, cross sections of fruits; E, I, seeds (X1). Fruits, Mossman and Dennis, southeastern coast of Florida.

(133)

Leguminosae

CASTANOSPERMUM AUSTRALE A. CUNN., MORETON BAY CHESTNUT

Disseminule: Fruit.
Description: Fruit (A) 15 to 20 cm long, about 5 cm wide, cylindrical, round in cross section, blackish brown to brown, surface smooth. Seed (B-D) 2.5 to 6 cm long, 2.5 to 4 cm wide, globose to ellipsoidal, round in cross section, dark brown, surface smooth and bearing a light colored linear hilum about as long as seed.
Buoyancy factor: Space around seeds and perhaps the seeds themselves.
Buoyancy: Not tested.
Viability: Not tested.
Currents: Pacific South Equatorial Current and perhaps other Pacific currents.

The duration and durability of drift Moreton Bay chestnut fruits and seeds are unknown. Our only record is from Ninety Mile Beach, North Island, New Zealand. This beach is known for receiving tropical debris (Vincent, 1957; Mason, 1961). The drift specimen is deposited at the Division of Scientific and Industrial Research, Botany Division, Christchurch, New Zealand. Moreton Bay chestnut, also known as black-bean tree or black-bean, is grown as an ornamental in southeastern Asia, South America, and southern California. Watt and Breyer-Branddwijk (1962) summarized the evidence concerning whether the seeds are poisonous or not. The evidence is conflicting, and to be safe neither the raw nor cooked seeds should be eaten. Care should be taken in handling the wood, because of the presence of poisonous saponins.

Leguminosae

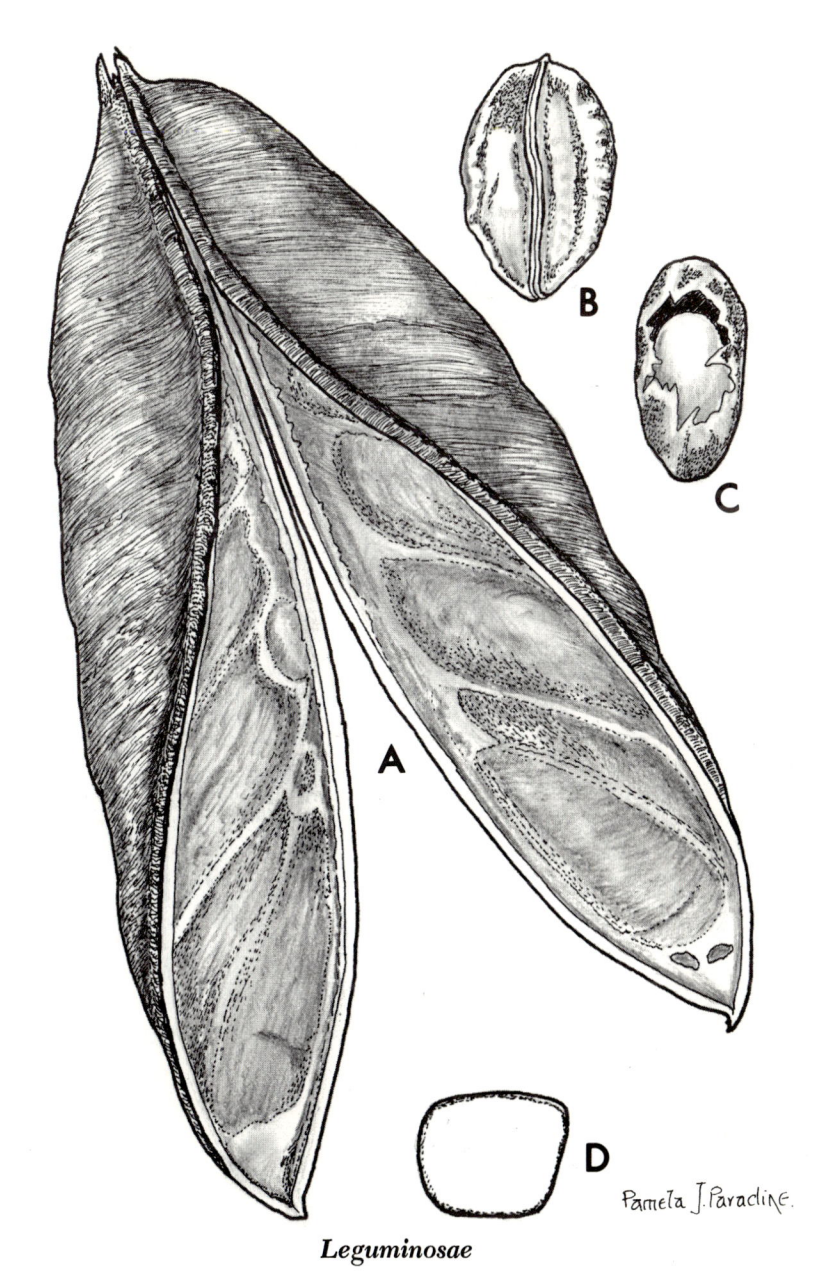

CASTANOSPERMUM AUSTRALIS A. CUNN.

Figure 52. *Castanospermum australe.* A, fruit; B-D, seeds. A-C, lateral views; D, cross section (X¾). Fruit and seeds, Ninety Mile Beach, North Island, New Zealand.

Leguminosae

DALBERGIA ECASTAPHYLLUM (L.) TAUB., COIN PLANT

DALBERGIA MONETARIA L. F., COIN PLANT

Synonym: *D. ecastaphyllum* is *D. brownei* (Jacq.) Urban in the drift literature.

Disseminule: Fruit.

Description: *D. ecastaphyllum* fruit (A-F) 20 to 30 mm long and wide, about 1 mm thick, ellipsoidal, strongly compressed in cross section, shiny to dull tan to reddish brown, surface smooth or nearly so. Embryo tips pointing away from each other (E). *D. monetaria* fruit (G-L) 25 to 40 mm long, 20 to 30 mm wide, 5 to 7 mm thick, ellipsoidal to globose, strongly compressed in cross section, shiny to dull brown to reddish brown, surface smooth or nearly so. Embryo tips pointing towards each other (K).

Buoyancy factor: Corky fruit coat and seed only partially filling seed cavity.

Buoyancy: *D. ecastaphyllum*—9 months; *D. monetaria*—not tested.

Viability: *D. ecastaphyllum*—about 25 percent viable; *D. monetaria*—not tested.

Currents: Tropical currents arising in the New World.

While the genus *Dalbergia* has a pantropic distribution, the two buoyant-fruit species are members of the tropical Atlantic New World flora. Within this region, *D. ecastaphyllum* is a frequent member of the mangrove swamp flora. *Dalbergia monetaria* is a rare species, having migrated by rivers from its interior home along northern South American rivers to the Atlantic coast and nearby islands. Because the latter species is less salt tolerant, there are fewer habitats which it can occupy. Thus its fruits are much less frequently stranded than are those of *D. ecastaphyllum*. Both species may be shrubby trees, shrubs, or vines. The wood of several *Dalbergia* species is known as rosewood and has wide use. Knife handles are made from cocobolo, another wood from this genus.

Leguminosae

DALBERGIA SPP.

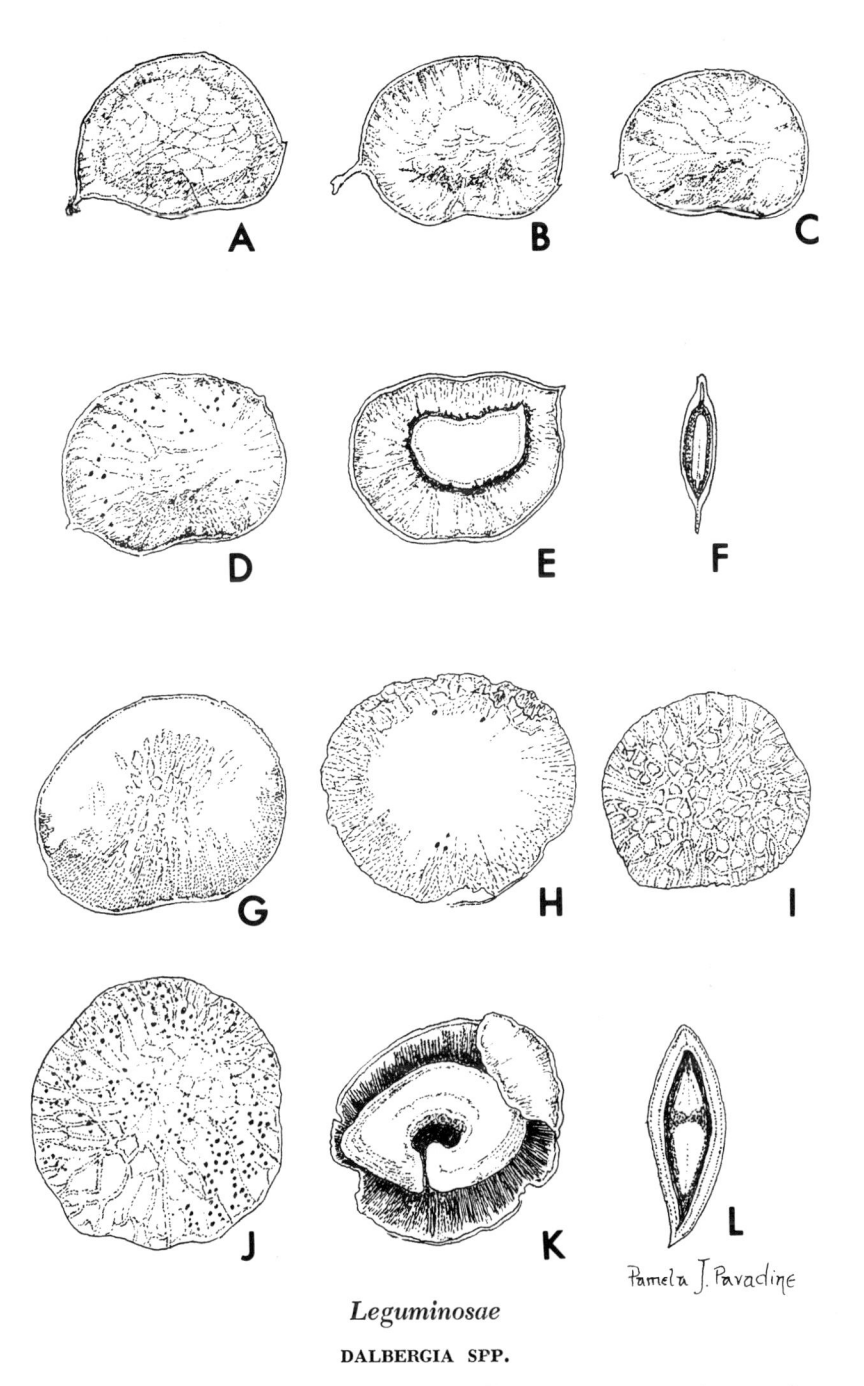

Figure 53. *Dalbergia ecastaphyllum* (A-F); *D. monetaria* (G-L). A-L, fruits. A-D, G-J, lateral views of fruits; E, K, fruits opened to expose seeds; F, L, cross sections of fruits (X1). Fruits, Gunn and Dennis, southeastern coast of Florida.

Leguminosae

DELONIX REGIA (HOOKER) RAF., ROYAL POINCIANA

Disseminule: One valve (one half of a fruit).
Description: Fruit (A, C) 40 to 60 cm long, 4 to 7 cm wide, strap shaped, strongly compressed in cross section, dark brown, outer surface nearly smooth (A), inner surface shallowly partitioned (C).
Buoyancy factor: Corky fruit.
Buoyancy: About 1 month.
Viability: No seeds present.
Currents: Tropical currents throughout the world.

Delonix regia, a beautiful and widely planted tree, is a native of Madagascar. It has been spread by man throughout the tropics and subtropics. Because it has brilliant showy scarlet-yellow flowers, the tree bears such common names as flame-tree, peacock-tree, and flamboyant-tree. The entire fruit is composed of 2 valves. The inner surface of each valve resembles a miniature step ladder, because of the shallow walls between each seed depression. There is no record that the entire fruit drifts. The individual drifting valves do not contain seeds (B). Therefore, drifting does not contribute to the spread of this species.

Leguminosae

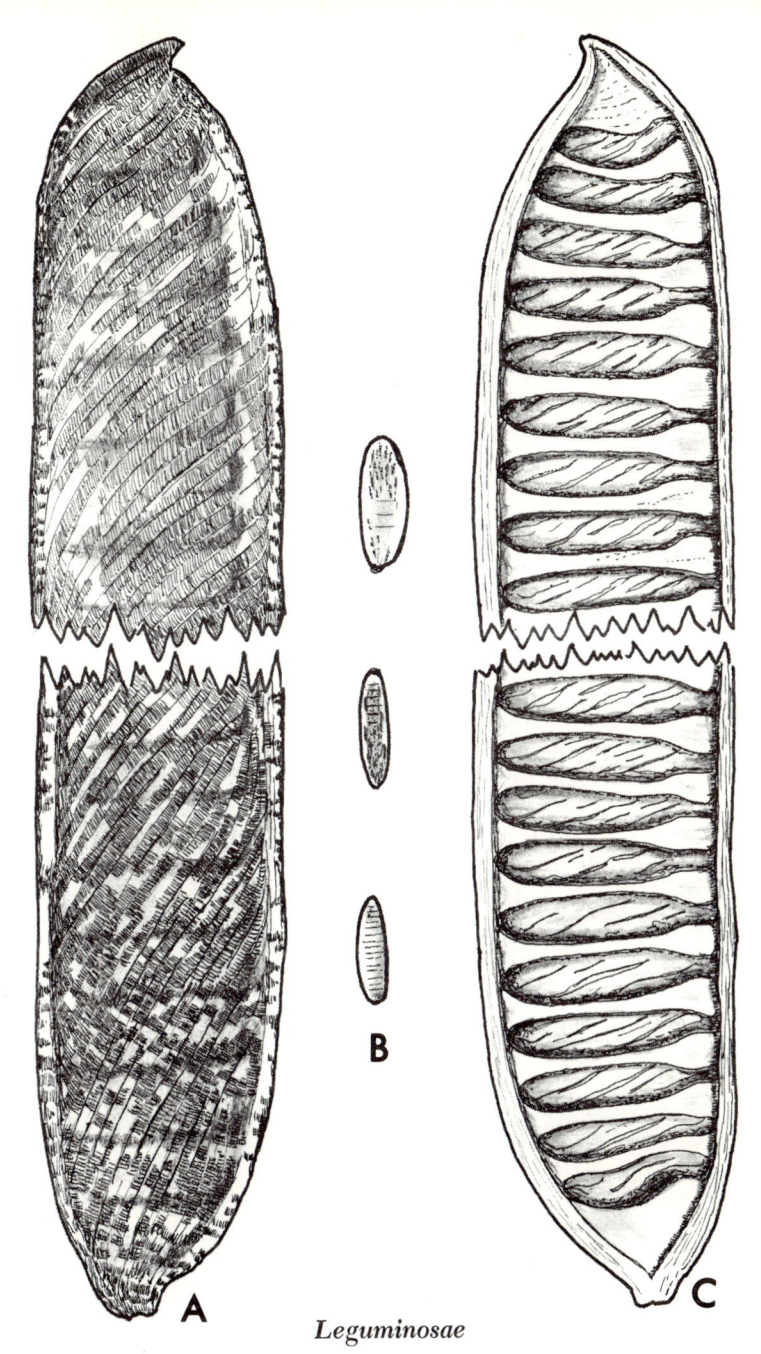

DELONIX REGIA (HOOKER) RAF.

Figure 54. *Delonix regia*. A, C, fruit valves (broken area indicates fruits too long to show entire length even at X¾); B, seeds. A, outer view; B, lateral views; C, inner view (X¾). Seeds, from tree; fruits, Mossman, beach at Palm Beach, Florida.

Leguminosae

DIOCLEA MEGACARPA ROLFE

DIOCLEA REFLEXA HOOKER F., SEA PURSE

DIOCLEA SP.

Desseminule: Seed.
Description: See species discussions.
Buoyancy: *D. reflexa*—at least 2 years; other species—not tested.
Viability: *D. reflexa*—most seeds viable; other species—not tested.
Currents: See species discussions.

Dioclea megacarpa seeds have been found stranded on beaches of Clipperton Island (Sachet, 1962), San Jose Island (Johnston, 1949), and Trinidad (Ridley, 1930). These sites are fairly close to the points or origin of the seeds, indicating that these seeds may not be as buoyant as the wide ranging seeds of *D. reflexa*. Seeds (A-C) 2.8 to 3 cm in diameter, globose, slightly compressed in cross section, lustrous monochrome light tan to brown, surface smooth. Hilum occupies about three-fourths of the seed circumference, and is about 5 mm wide at each end and 3 mm wide in the middle.

Dioclea panamensis Walpers (not shown), a native vine of Panama to Ecuador, may be the source of seeds collected by Guppy (1917) in the estuary of the Guayaquil River, Ecuador.

Dioclea reflexa, sea purse or saddle-bean, is the only species in the genus to have achieved a pantropic distribution by drifting from its Asian home. These seeds are also one of the few tropical disseminules which have reached northern Europe via the Gulf Stream and Tristan da Cunha by the Brazil Current. Some of these seeds may be from plants of *D. violacea* Bentham or *D. wilsonii* Standley according to Richard Maxwell, Indiana University Southeast, New Albany. Seeds (D-K) are 2.5 to 3.5 cm long, 2 to 2.5 cm wide, round with one flattened side, strongly compressed (I) to slightly compressed (K) in cross section, lustrous monochrome tan to dark brown or mottled, surface smooth to wrinkled. Hilum occupies about three-fourths of the seed circumference and is about 2 mm wide.

Dioclea sp. seeds (L-N) from Canton Island beaches are 3 to 4 cm long, 2.5 to 3 cm wide, usually round in outline, strongly compressed in cross section, lustrous black, surface smooth to wrinkled and bearing a conspicuous bulge near the center. Hilum occupies about three-fourths of the seed circumference and is about 2 mm wide.

Leguminosae

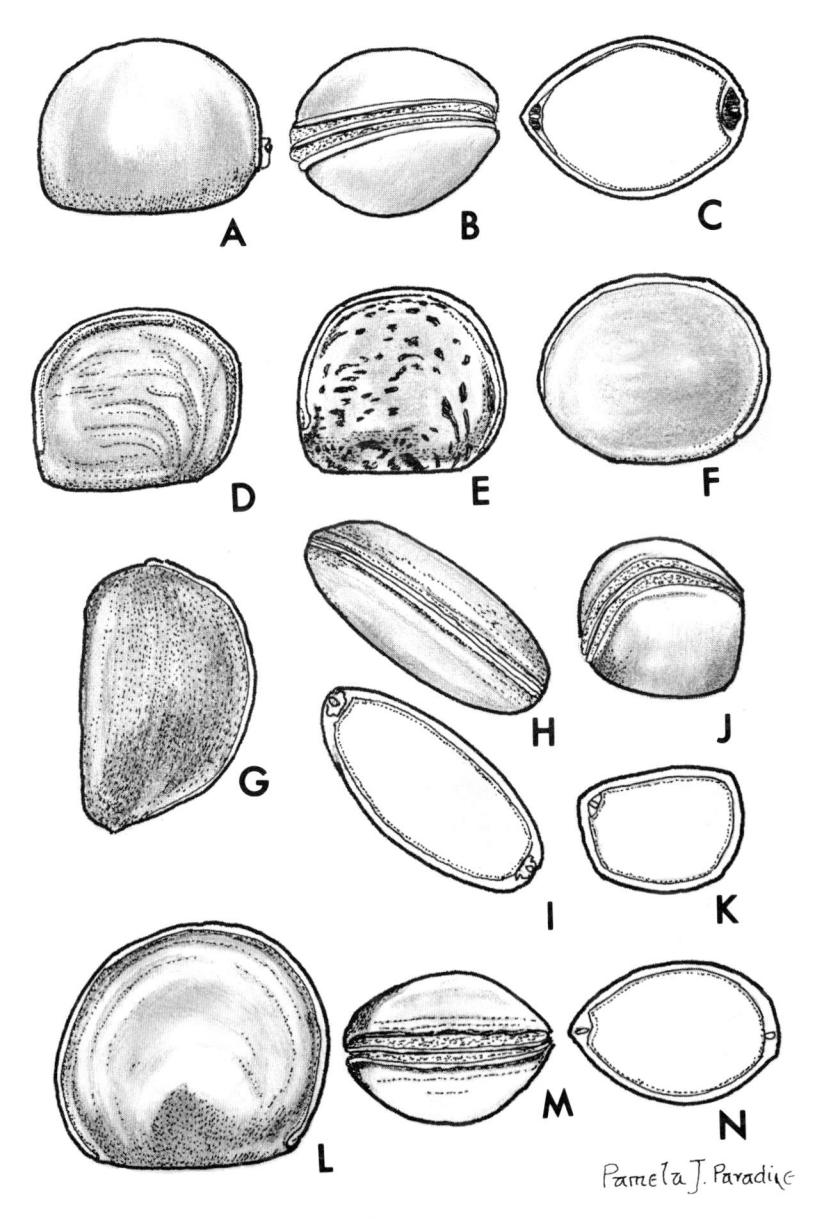

DIOCLEA SPP.

Figure 55. *Dioclea megacarpa* (A-C); *D. reflexa* (D-K); *D. sp.* (L-N). A-N, seeds. A, D-G, L, lateral views; B, H, M, hilar views; C, I, K, N, cross section (X1). *D. megacarpa*, from vine; *D. reflexa*, Gunn and Dennis, southeastern coast of Florida; *D. sp.*, Degener and Degener, beaches of Canton Island.

(141)

Leguminosae

ENTADA GIGAS (L.) F. & R., SEA HEART

Synonym: *E. scandens* Bentham in the New World drift literature.
Disseminule: Seed, rarely fruit compartment.
Description: Seed (B-F) 4 to 6 cm in diameter, 1.5 to 2 cm thick, cordate (B, C) or seldom oblong (D) to ellipsoidal (E), strongly compressed in cross section, dull to somewhat lustrous chocolate to mahogany brown, surface smooth or nearly so.
Buoyancy factor: Intercotyledonary cavity.
Buoyancy: At least 2 years.
Viability: Most seeds viable.
Currents: Tropical currents arising in the New World.

According to the drift literature, *E. scandens* was a pantropic species whose distribution was derived from drifting. The species concept of *E. scandens* has been abandoned in favor of a New World species, *E. gigas*, and at least one Old World species, *E. phaseoloides* (L.) Merrill. Either *E. phaseoloides* gave rise to *E. gigas*, or both had a common ancestor. *Entada gigas* is a high climbing woody tropical vine whose seeds have reached Norway (Lindman, 1882) from the Caribbean region via the Gulf Stream. Viable seeds and seeds whose viability has not been ascertained have been stranded on intervening beaches: Denmark (Erslev, 1877), Orkneys (Wallace, 1700), Scotland (Shackleton, 1959, Sibbald, 1694; Sloane, 1707), Hebrides (Necker de Saussure, 1821), Ireland and Northern Ireland (Colgan, 1919), England (Hobbs, 1969; Lloydd-Jones, 1898), Iceland (Shackleton, 1959), Greenland (Kohl, 1868), Massachusetts and Carolinas (Gunn and Dennis, 1972b), and Florida (Gunn, 1968). Perhaps the most interesting reference is Sernander (1901) who reported that sea hearts were found in Swedish post-glacial peat bogs in a semi-fossil state. A few seeds which bear imprints of teeth have been collected from southeastern Florida beaches.

Fruits of *Entada* are among the longest fruits. They are usually 1 to 2 meters long and 8 to 12 cm broad. Unlike the fruits of other species in the genus, sea heart fruits are not woody and are unable to support their own weight when held horizontally. The twisted fruits contain 10 to 15 seeds, each in an individual compartment. These compartments are fragile, and they are only occasionally stranded (A).

Leguminosae

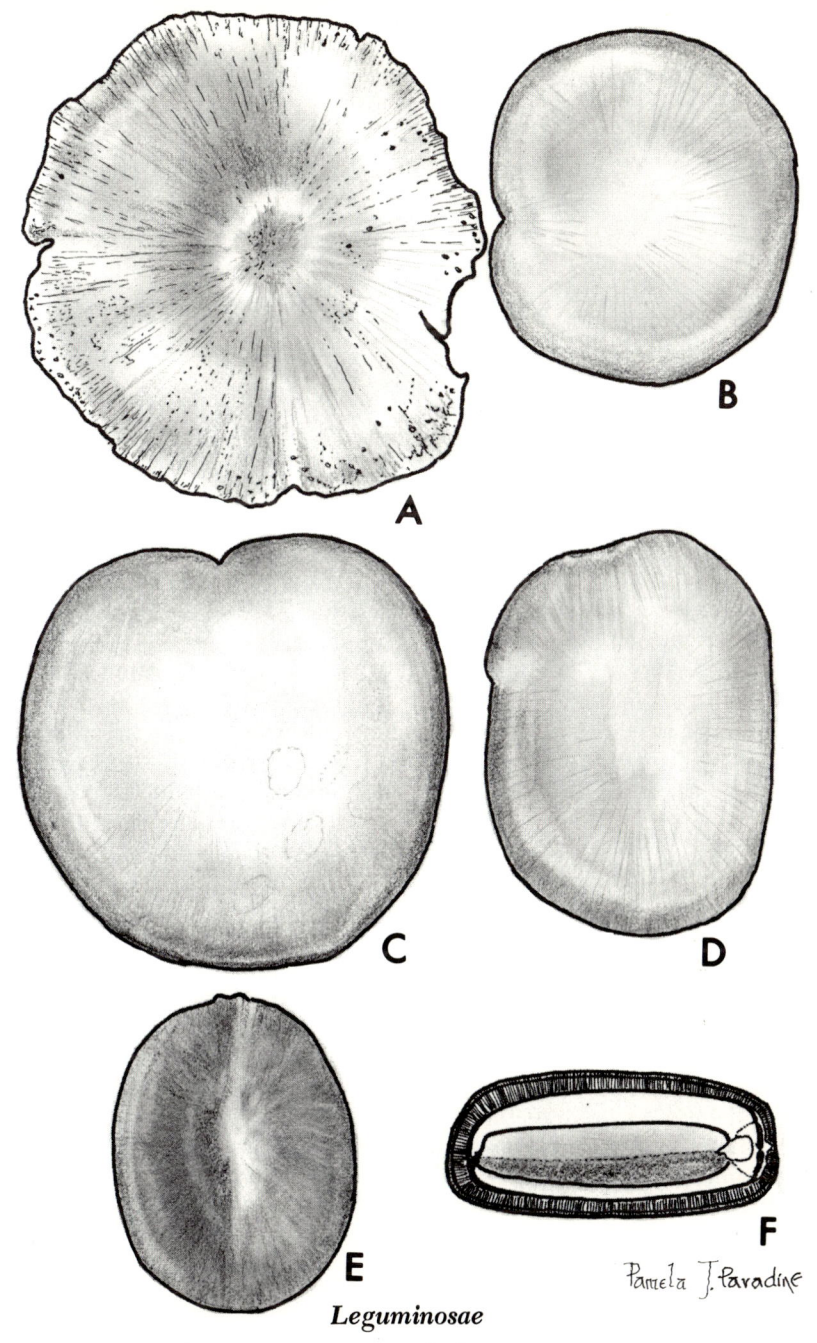

ENTADA GIGAS (L.) F. & R.

Figure 56. *Entada gigas*. A, fruit compartment; B-F, seeds. A-E, lateral views; F, cross section (X1). Fruit and seeds, Gunn and Dennis, southeastern coast of Florida.

Leguminosae

ENTADA PHASEOLOIDES (L.) MERRILL, SNUFFBOX SEA-BEAN

Synonym: *E. scandens* Bentham is the Old World drift literature.
Disseminule: Seed or fruit segment.
Description: Seed (B-F) 3 to 6 cm long, 2.5 to 5 cm wide, up to 2 cm thick, rectangular (B, D) to round (C, E), strongly compressed in cross section, dull to somewhat lustrous chocolate to mahogany brown, surface smooth or nearly so.
Buoyancy factor: Intercotyledonary cavity.
Buoyancy: Not tested, though literature indicates excellent buoyancy.
Viability: Not tested, though literature indicates most viable.
Currents: Tropical currents arising in southeast Asia and the Pacific region.

The seeds of *E. phaseoloides* and *E. gigas* (see preceding pages) are similar. While some seeds of these two species cannot be separated, generally seeds of *E. phaseoloides* are longer than wide and bear more or less straight sides (B, D). Seeds of *E. gigas* are generally as long as wide and have a curved outline. Seeds of *E. phaseoloides* are found on beaches from southeast Africa to Hawaii. We cannot be certain that all of the *Entada* drift seeds in the Africa-Oceania region are *E. phaseoloides*, though most are. Buoyant seeds of related species include *E. formosana* Kanehira, *E. gogo* (Blanco) Johnston, and *E. koshunensis* Hay. & Kanehira.

Entada phaseoloides is a high climbing woody vine whose fruits are woody and sturdy. The fruit is capable of supporting its own weight when held horizontally. Stafford (1905) aptly compared the fruit to the scabbard of a sword. Each seed is enclosed within an individual compartment, and this compartment is fragile. Occasionally a compartment may be stranded (A). Muir (1937) reported "extensive lacerations probably caused by a fish" on a snuffbox sea-bean seed. Like the seeds of the sea heart (*E. gigas*), these seeds have myriad uses. Entire seeds are used in games, as baby teethers, and when hollowed out and hinged as snuff, tinder, and match boxes. Ground seeds are used externally to relieve inflammation. Seeds are also used as contraceptives, purgatives, coffee substitutes, and even as food.

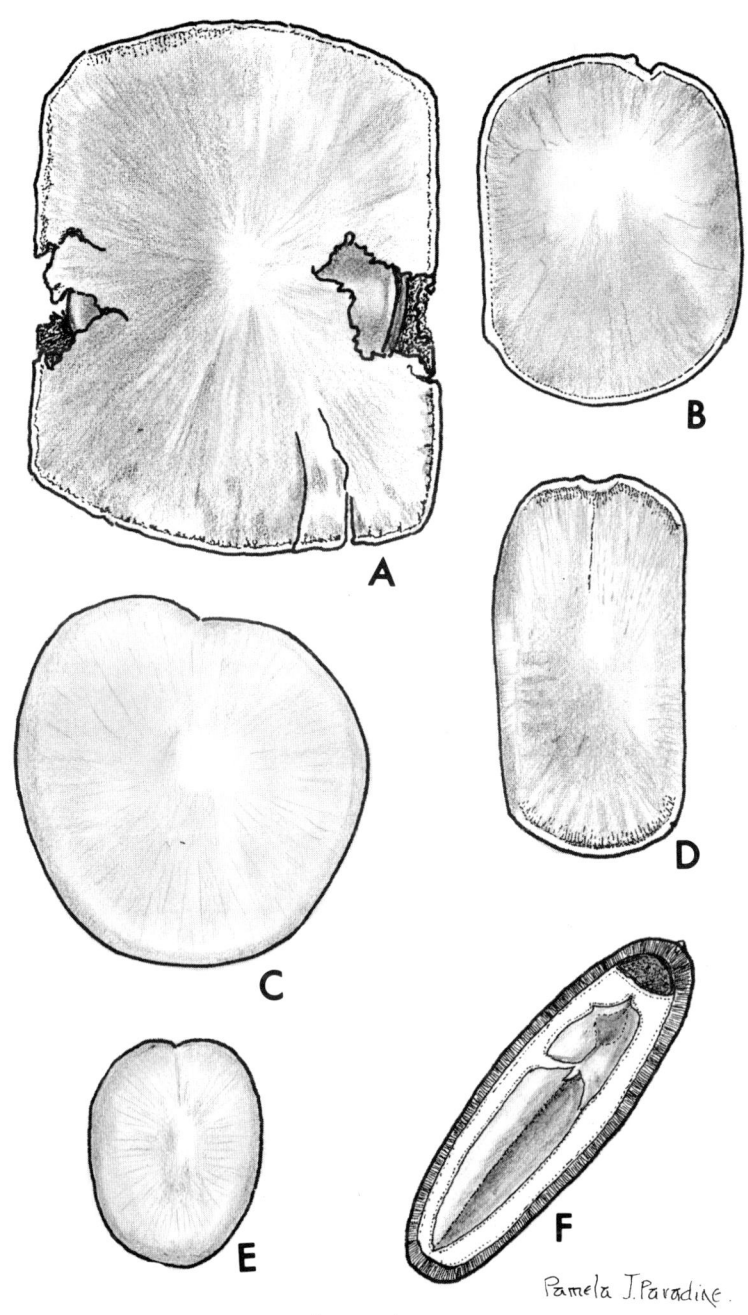

Leguminosae

ENTADA PHASEOLOIDES (L.) MERRILL

Figure 57. *Entada phaseoloides*. A, fruit segment; B-F, seeds. A-E, lateral views; F, cross section (X1). Fruit, Clocker, beaches of Viti Levu, Fiji; seeds, Degener and Degener, beaches of Canton Island.

Leguminosae

ENTEROLOBIUM CYCLOCARPUM (JACQ.) GRISEB., LARGE EAR POD
ENTEROLOBIUM TIMBOUVA MARTIUS, SMALL EAR POD

Disseminule: Fruit or fruit segment.
Description: *E. cyclocarpum* fruit (A-C) up to 11 cm in diameter, ends often overlapping, forming a complete circle, strongly compressed in cross section, dull to lustrous chocolate brown, surface smooth except for lumps formed by seeds. *E. timbouva* fruit (E-F) 5 to 6 cm in diameter, ends coming close together but not overlapping, strongly compressed in cross section, dull to lustrous blackish brown, surface smooth. Fruits of both species resemble the human ear.
Buoyancy factor: Lightweight fruit tissues and space around seeds.
Buoyancy: *E. cyclocarpum*—about 1 month; *E. timbouva*—not tested.
Viability: Not tested.
Currents: Tropical currents arising in the New World.

Enterolobium cyclocarpum is a large tree (up to 30 meters tall) that is native to Central and tropical South America. It has been spread by man to most tropical countries as an ornamental. The tree is an inland species, possessing little salt tolerance. Fruits are carried by rivers to the ocean where they may drift. Ocean currents have played little role in the spread of this species. Fruits are so thin that the outline of the 11 to 20 seeds per fruit can be seen on the surface of the fruit. Because of their thinness, the fruits are usually stranded in a damaged condition.

Enterolobium timbouva fruits are smaller and thicker and are usually stranded whole. They bear no impression of the seeds on their surfaces. This New World tree has also been planted, but not to the extent of the other species. The seeds of neither species float. Both seeds (D, G) bear a light-colored, closed line on each face.

Leguminosae

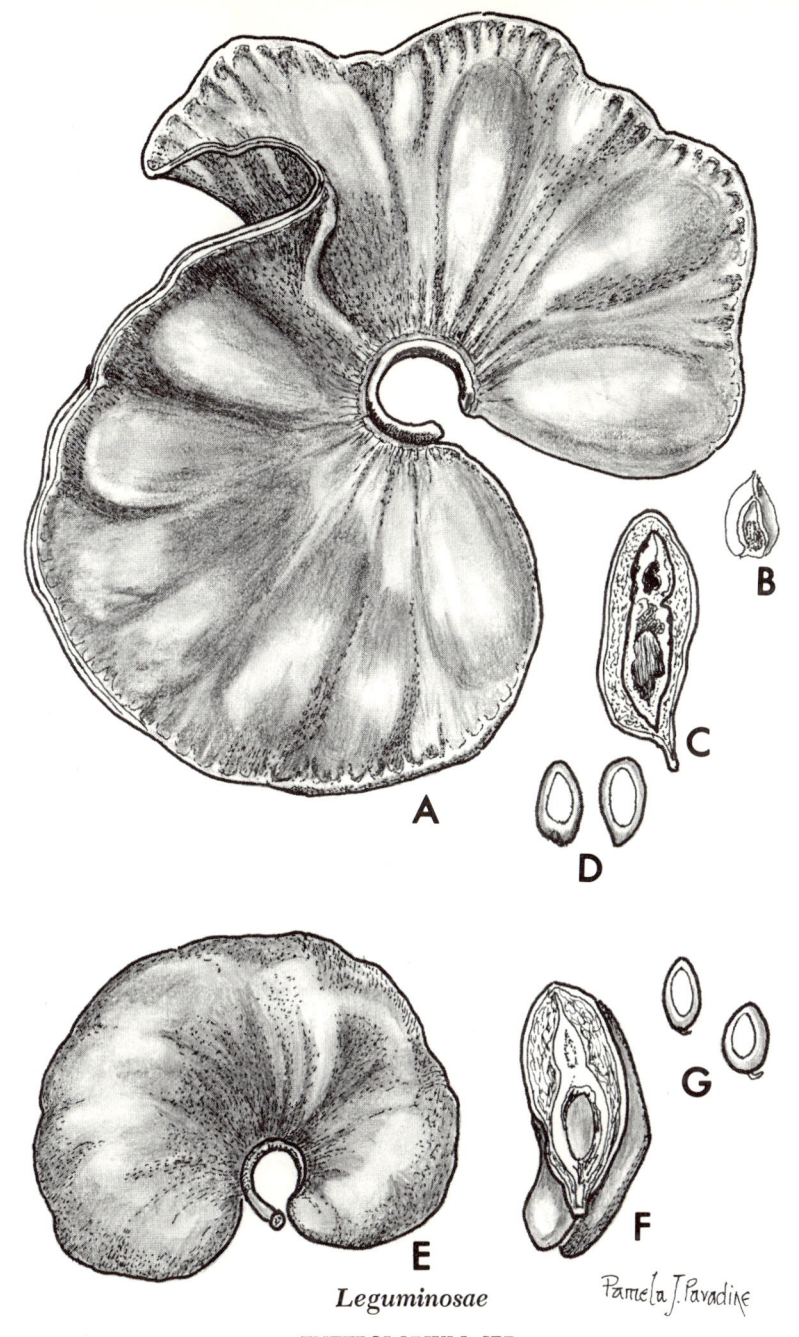

Pamela J. Paradine

ENTEROLOBIUM SPP.

Figure 58. *Enterolobium cyclocarpum* (A-D); *E. timbouva* (E-G). A, C, E-F, fruits; B, dried fruit pulp around seed; D-G, seeds. A, B, D, E, G, lateral views; C, F, cross sections (X1). Fruits, Mossman and Dennis, southeastern coast of Florida (seeds extracted from drift fruits).

Leguminosae

ERYTHRINA SPP. CORALBEAN

Disseminule: Seed.
Description: Seed (A-F, H-O) 1 to 1.8 cm long, oblong, round or nearly so in cross section, lustrous scarlet, vermillion, orange-red, yellow, purplish, brown, or black, surface smooth. Hilum conspicuous, dark colored, ellipsoidal to ovate, about half the seed length.
Buoyancy factor: Buoyant cotyledonary tissue.
Buoyancy: About 15 months.
Viability: Most seeds viable.
Currents: Tropical currents arising in the New and Old World.

Species in this genus are trees or shrubs which are often prickly. The 200 species in the genus are native to the New and Old World tropics and subtropics. Guppy (1917) tried to make a case for long distance drift of *Erythrina* seeds by citing a drawing in Wallace (1693) and the word "bluesteen" (bent-stone) in old Scandinavian drift literature. While Guppy thought that the drawing and common name referred to an *Erythrina* seed, Linnaeus (1789) regarded bluesteen as a common name for *Piscidia erythrina* L. We believe that these are not references to *Erythrina* seeds, and we doubt that the Scandinavian bluesteen is a *Piscidia*. No *Erythrina* seed has been reported from any beach north of Florida, even though some seeds are brightly colored and thus conspicuous. No drift *Piscidia* seed has been reported.

We have been unable to identify drift seeds to species, primarily because we lack authentic comparison seed samples. Seeds of *E. variegata* L. var. *orientalis* (L.) Merrill have been found on beaches throughout the world, viz., southeastern Florida; Quintana Roo, Mexico; Riversdale, South Africa; Ujae Atoll, Marshall Islands; Viti Levu, Fiji Islands; and Canton Island. The plant is native of southeastern Asia and the Pacific region. It has been spread by man to the New World tropics. The seeds are about 1.5 cm long, brown to brownish purple.

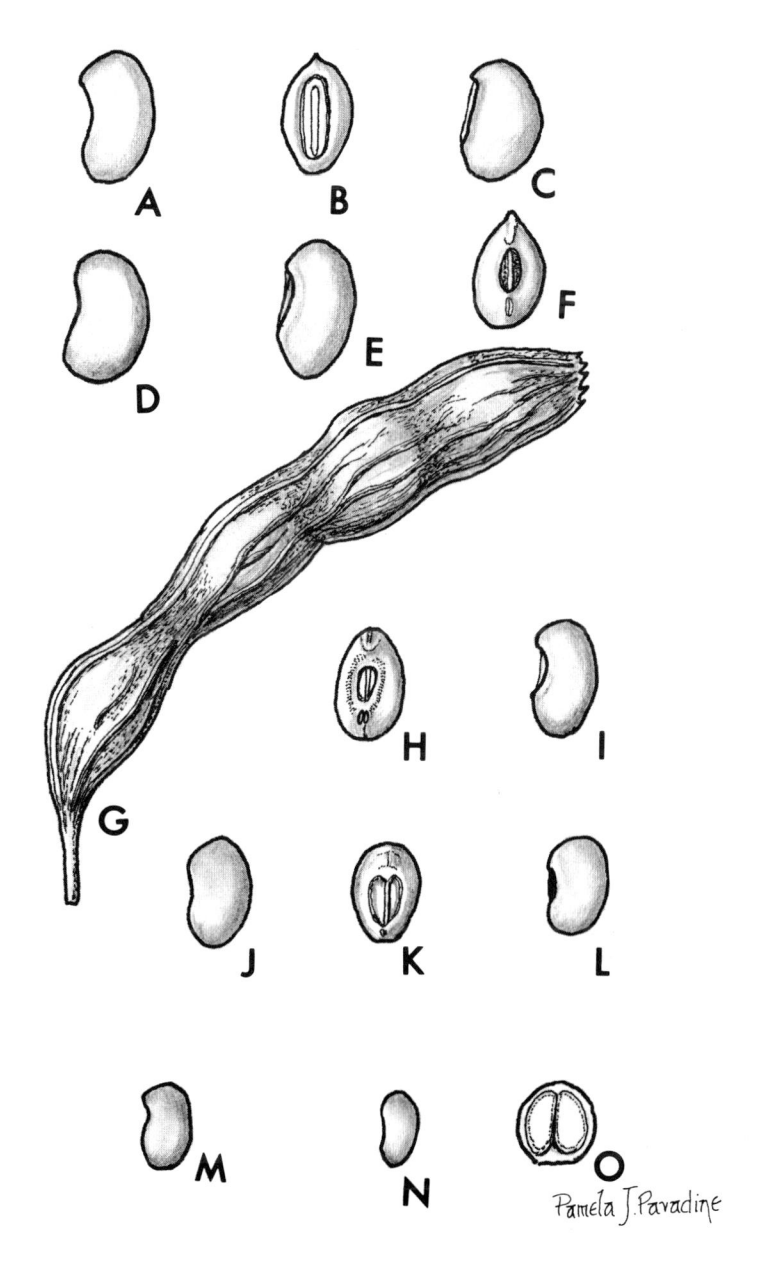

Leguminosae

ERYTHRINA SPP.

Figure 59. *Erythrina* spp. A-F, H-G, seeds; G, fruit. A, C, D-E, G, I, J, L-N, lateral views; B, F, H, K, hilar views; O, cross section (X1). Fruit, from plant; seeds, Gunn and Dennis, southeastern coast of Florida.

(149)

Leguminosae

HYMENAEA COURBARIL L., WEST INDIAN LOCUST

Disseminule: Fruit.
Description: Fruit (A-B) 5 to 14 (rarely 20) cm long, 3.5 to 5 (rarely 10) cm wide, 2.5 to 4 cm thick, oblong, slightly compressed in cross section, dark brown, surface slightly rough.
Buoyancy factor: Empty space in seed cavity.
Buoyancy: At least 2 years.
Viability: Not tested. Viable seeds were recovered from a fruit stranded on a Martha's Vineyard, Massachusetts beach in 1961 (Main • lines, 1971).
Currents: Tropical currents arising in the New World and perhaps in the Old World.

This small to large tree has been spread by its fruits drifting throughout the American tropics. It has been introduced to the Old World tropics by man, because there is no New World to Old World transport current. The tree inhabits low dry forests, and the fruits are carried by rivers to the ocean. The report of *Hymenaea* seeds beached on the Irish coast (Blake, 1825) is not correct. The seeds do not float. Because Blake described seeds of *Caesalpinia bonduc, Merremia discoidesperma,* and *Mucuna* spp., but did not mention *Entada gigas,* it is reasonable to assume that the *Hymenaea* seed reference should be applied to *E. gigas.* While the fruits of *H. courbaril* have not been reported as reaching northern Europe, they have reached Massachusetts intact. The indehiscent fruit is quite durable. The walls are 3 to 4 mm thick. The partially empty fruit contains 2 to 6 seeds which may be embedded in a thick mealy pulp. The seeds (C-D) are 2 to 3 cm long and dark reddish brown. The gummy sap which exudes from the trunk of the tree is used to make varnish and as an incense.

Leguminosae

HYMENAEA COURBARIL L.

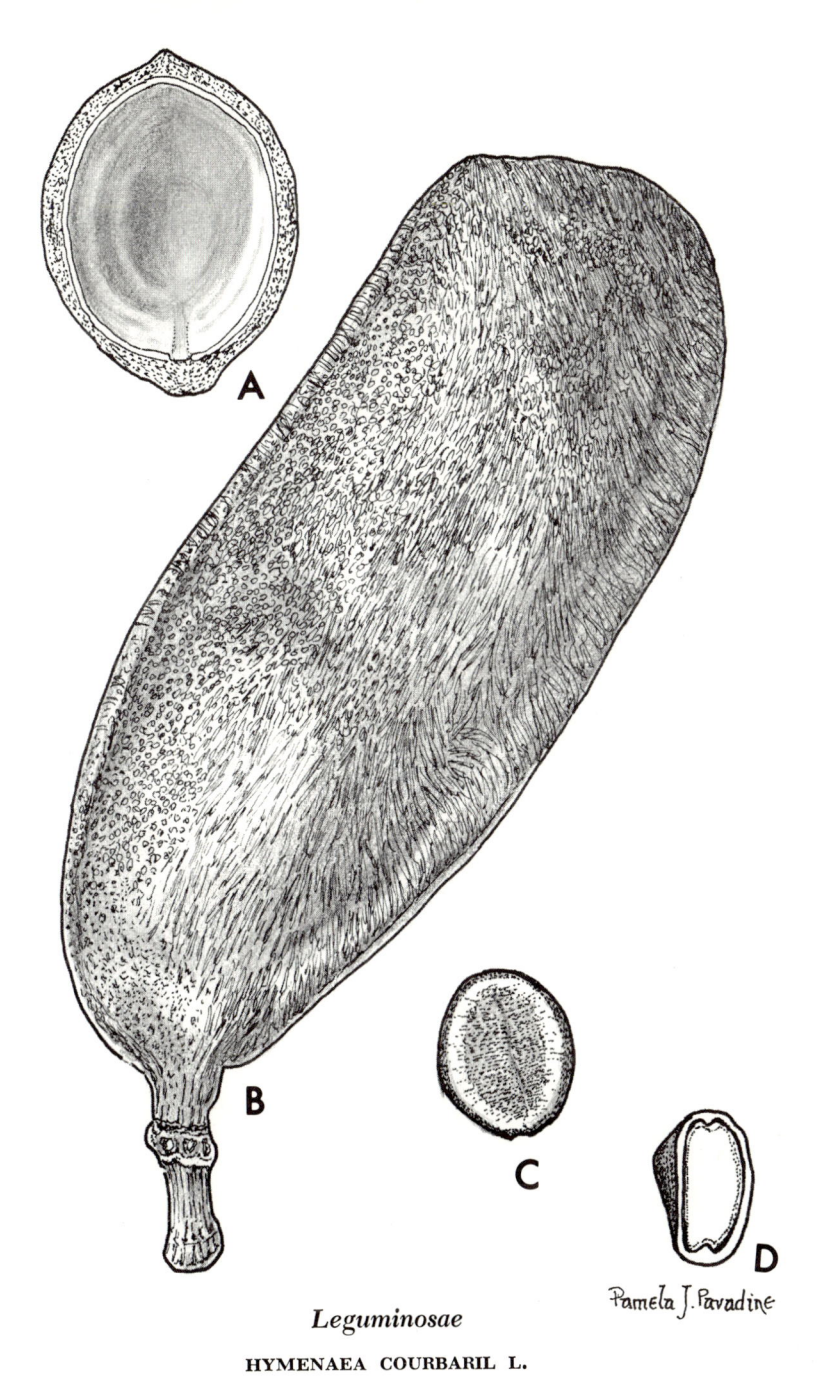

Figure 60. *Hymenaea courbaril.* A-B, fruits; C-D, seeds. A, D, cross sections; B-C, lateral views (X1). Fruits, Gunn and Dennis, southeastern coast of Florida (seeds extracted from drift fruits).

Leguminosae

INTSIA BIJUGA (COLEBR.) O. KUNTZE

Synonym: *Afzelia bijuga* (Colebr.) A. Gray in the drift literature.
Disseminule: Seed.
Description: Seed (A-L) 2.5 to 4 cm long, 1.5 to 3.5 cm wide, 4 to 6 mm thick, irregularly ovate, cordate or elongate, strongly compressed in cross section, dull to lustrous black to blackish brown, surface nearly smooth, marked with faint concentric fracture lines.
Buoyancy factor: Buoyant cotyledonary tissue.
Buoyancy: Not tested.
Viability: Not tested. Muir (1930) germinated stranded seeds.
Currents: Tropical currents arising in the southeastern Asia and Pacific regions.

Intsia bijuga has a range from Madagascar to Fiji, and this range is the result of drifting. Viable seeds have entered the Atlantic Ocean after rounding the tip of South Africa. However, the species has not achieved a pantropic distribution, indicating that the seeds have either lost their viability or have become stranded in habitats unsuitable for germination or survival. The tree grows on sand and coral beaches. If found inland, the tree grows in areas which are periodically inundated. Like other species which occupy littoral and inland habitats, Guppy (1917) made a dubious case for buoyant littoral seeds and non-buoyant inland seeds. Our thoughts on this subject are summarized in Chapter 1. Muir (1930) reported this seed was the third most common drift disseminule on beaches of southern South Africa. Most of his seeds were covered with polyzoa skeletons indicating that the seeds had been in the ocean water for at least several weeks. Because the tree is unknown in South Africa, Muir thought that the source of the drift seeds was Madagascar, the nearest place where native trees grow. No one has recorded whether the fruits float or not. The fruit is 8.5 to 23 cm long, 4 to 8 cm wide, linear to oblong, flat, and bears 1 to 8 seeds. Each seed is borne within an individual compartment. Because the fruits dehisce tardily, one would expect them to float.

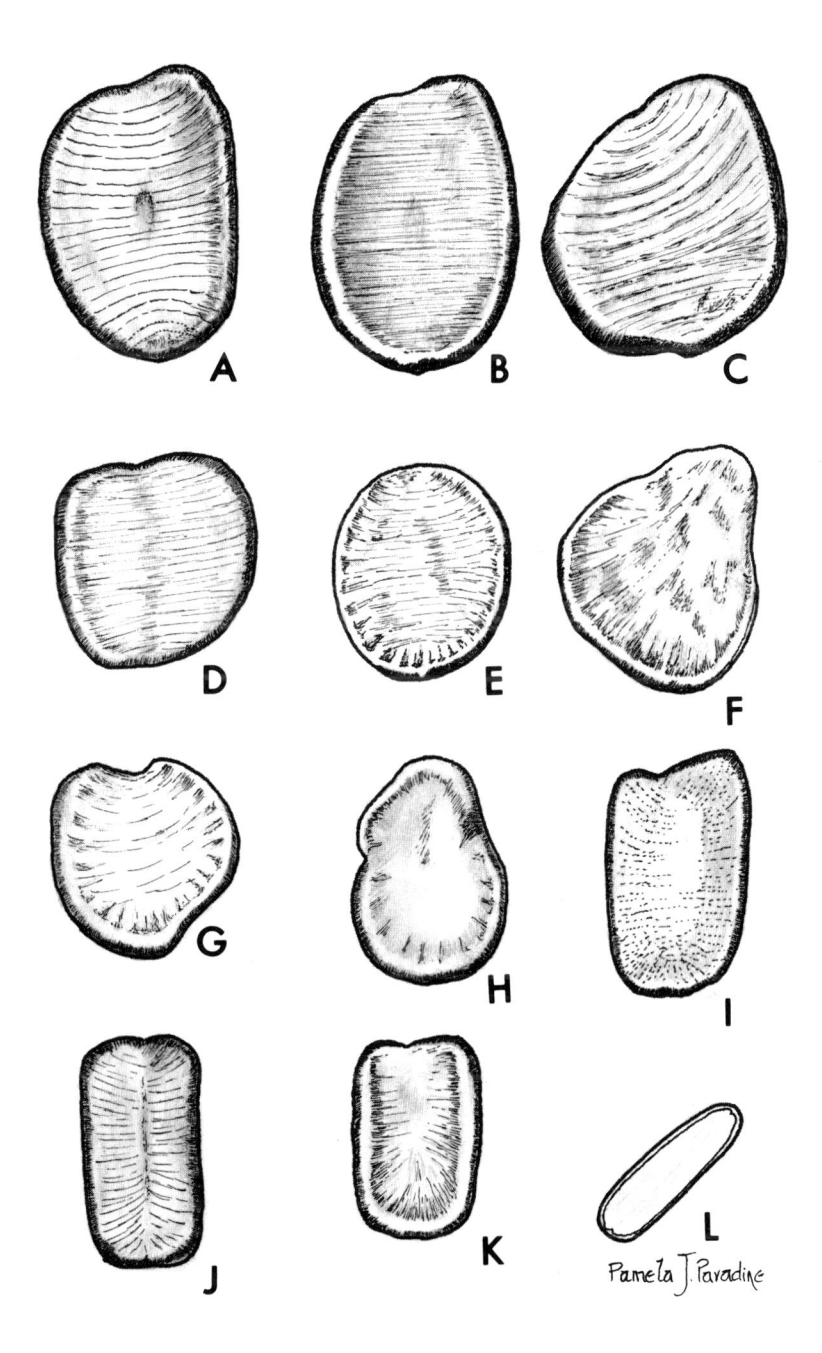

Leguminosae

INTSIA BIJUGA (COLEBR.) O. KUNTZE

Figure 61. *Intsia bijuga* seeds. A-K, lateral views; L, cross section (X1). Seeds, Degener and Degener, beaches of Canton Island.

Leguminosae

MORA EXCELSA BENTHAM, MORA

Synonym: *Dimorphandra mora* Bentham and Hooker f. in the drift literature.

Disseminule: Seed or half seed.

Description: Seed (A, B, D) about 8 cm long, 3 to 5 cm wide, oblong with one side deeply notched about midway along the length, compressed in cross section, dark brown to blackish brown, outer surface rough (seed coat often exfoliating), inner surface light brown, smooth. Hilum recessed in the conspicuous medial notch. Each cotyledon (C) about 6 mm thick.

Buoyancy factor: Intercotyledonary cavity; half cotyledons do not float.

Buoyancy: About 6 months.

Viability: None viable.

Currents: Tropical currents arising in the New World.

Mora is a large tree native to northeastern South America and Trinidad, where it usually grows in swamps or on ground which is periodically flooded. Because the trees are so large and the ground is usually saturated, the trees have a well-developed fluted base which lends stability. The buoyant seeds are carried to the ocean by rivers. These seeds closely resemble *Mora* sp. (Fig. 63 B-D). The seeds of *Mora* sp. are notched or dimpled near one end, are less than 7 cm long, and have a reddish brown to black inner cotyledonary surface. While *M. excelsa* seeds weigh less than those of *M.* sp., they are relatively poor drifters. Most seeds and all half seeds roll ashore with the tide, like sea shells. The seeds of *M. excelsa* are second in size to the seeds of *M. oleifera* (Fig. 63 A).

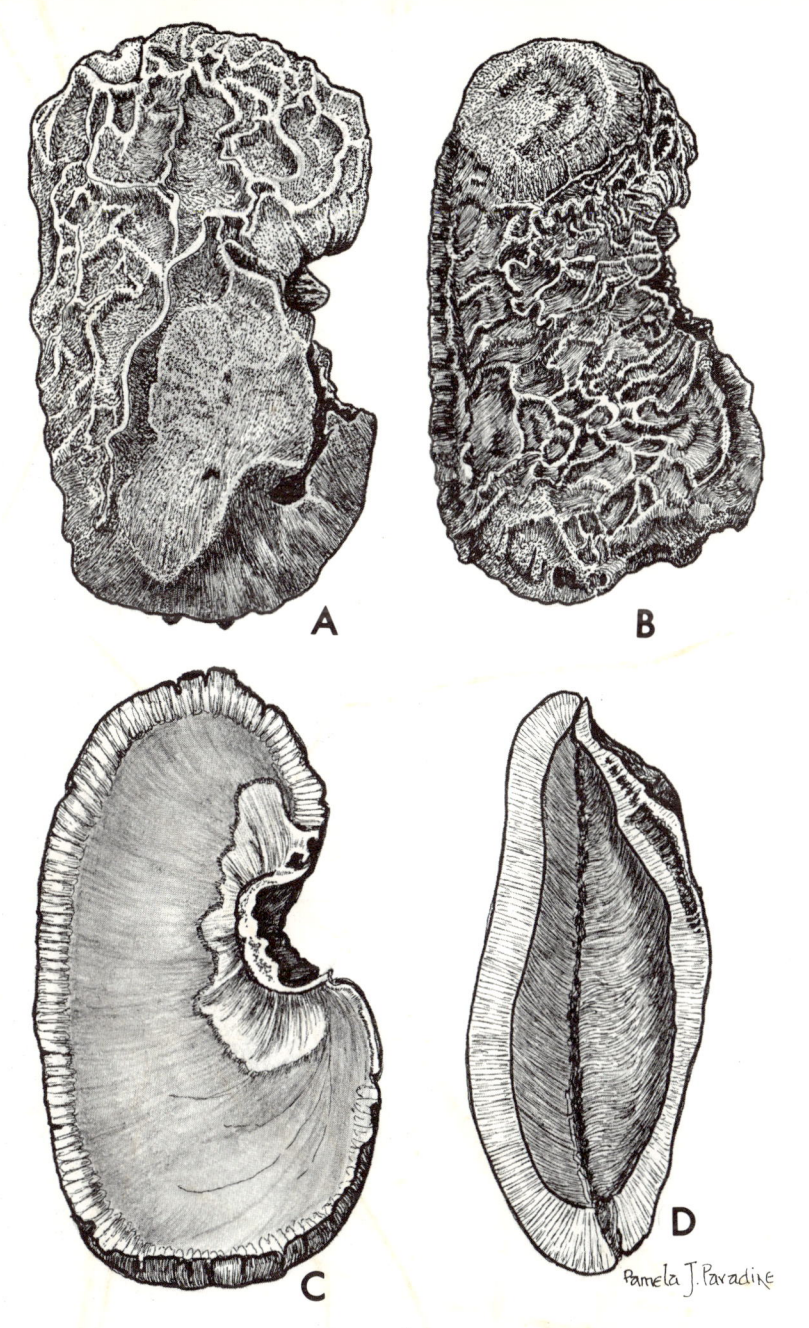

Leguminosae

MORA EXCELSA BENTHAM, MORA

Figure 62. *Mora excelsa* seeds. A-B, lateral views; C, half seed; D, longitudinal section (X1). Seeds, Gunn and Dennis, southeastern coast of Florida.

(155)

Leguminosae

MORA OLEIFERA (TRIANA) DUCKE
MORA SP.

Synonyms: *Mora* is *Dimorphandra* in some drift literature; *M. oleifera* is *D. megistosperma* Pittier.

Disseminule: *M. oleifera*—seed; *M.* sp.—1 cotyledon, rarely seed.

Description: *M. oleifera* seed (A) 12 to 18 cm long, 8 to 12 cm wide, to 8 cm thick, ellipsoidal with a deep lateral notch, round in cross section, dark brown, surface nearly smooth. Hilum located at base of notch. *M.* sp. seed (B) or single cotyledon (C-D) 4.5 to 7 cm long, 2 to 3 cm wide, each cotyledon 8 to 10 mm thick, oblong with a subterminal notch or dimple, seed round in cross section, dark brown, outer surface convex and rough, inner surface concave, smooth, and dark brown. Hilum located at base of notch or dimple.

Buoyancy factor: Both species—intercotyledonary cavity; single cotyledons of *M.* sp. roll ashore.

Buoyancy: Not tested.

Viability: Not tested. No single cotyledons of *M.* sp. bore an embryonic axis; therefore, none could be viable.

Currents: *M. oleifera*—tropical currents along the Pacific coast of northern South America and Central America; *M.* sp.—tropical currents arising in the New World.

While seeds of *M. oleifera* are exceeded in size by the endocarps and seeds of the coco-de-mer (*Lodoicea maldivica*) and the coconut (*Cocos nucifera*), they do possess the largest embryo of any seed. Most palm seeds, including the coco-de-mer and the coconut, are mainly composed of endosperm. Palm embryos are quite small. The seeds of *M. oleifera* often drift without their seed coat. Unlike *M. excelsa* and *M.* sp., the cotyledons remain completely and permanently united, insuring that the seed will drift. The trees, which may attain a height of 25 m, inhabit tidal estuaries along the Pacific side of tropical America, from Costa Rica to western Colombia. This unusual seed is seldom collected, because beaches along this coast are seldom visited by collectors. In addition seeds often germinate soon after falling into the mud or water below the parent tree. The fruits are not unusually large. They may be 1- or 2-seeded and seldom are more than 25 cm long and 13 cm wide.

Seeds of *M.* sp. closely resemble *M. excelsa* (Fig. 62). For a comparison of the seeds, see the preceding discussion of *M. excelsa*. Like *M. excelsa* the single cotyledons roll ashore with the tide, as do sea shells which they closely resemble.

Leguminosae

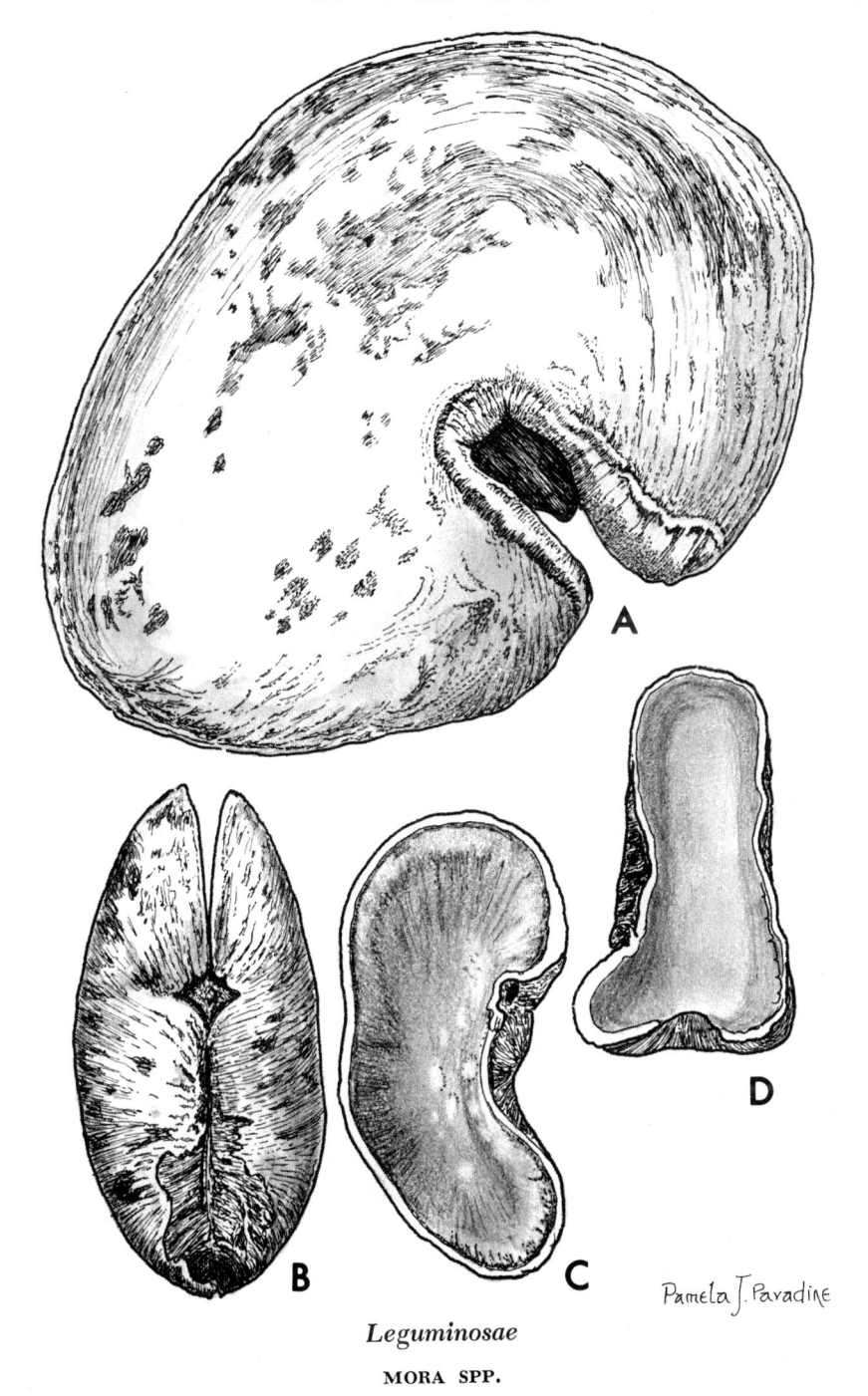

MORA SPP.

Figure 63. *Mora oleifera* (A); *Mora* sp. (B-D). A-D, seeds. A, lateral views; B, hilar view; C-D, half seeds (X1). *M. oleifera*, from tree; *M.* sp., Gunn and Dennis, southeastern coast of Florida.

Leguminosae

MUCUNA FAWCETTII URBAN, TRUE SEA-BEAN

MUCUNA GIGANTEA (WILLD.) DC.

MUCUNA NIGRICANS STEUDEL

MUCUNA SLOANEI F. & R., TRUE SEA-BEAN

MUCUNA URENS (L.) MEDIKUS, TRUE SEA-BEAN

These *Mucuna* spp. are vines, and they produce buoyant seeds which bear a hilum that occupies at least ¾ of the seed circumference. New World species include *M. fawcettii, M. sloanei,* and *M. urens.* These seeds may be carried alive by the Gulf Stream System to northern Europe. They are collectively known as true sea-beans, burning-beans, and horse eye-beans. Old World species include *M. gigantea, M. nigricans,* and *M.* spp., all from beaches of Canton Island.

M. fawcettii seed (Fig. 64 A-F) 25 to 35 mm in diameter, 20 to 25 mm thick, round, compressed in cross section, lustrous blackish brown to dark brown with a yellowish border along the hilum, surface essentially smooth. Hilum black, about 10 mm broad. Intercotyledonary cavity buoys seeds for more than 1.5 years.

M. sloanei seed (Fig. 64 G-L) 20 to 40 mm in diameter, 15 to 20 mm thick, round, compressed (rarely nearly round) in cross section, lustrous monochrome reddish brown, brown or blackish brown or mottled with black, a well to poorly defined (rarely absent) yellowish border along hilum, surface essentially smooth. Hilum black, 4 to 6 mm broad. Intercotyledonary cavity buoys seeds for more than 1.5 years.

M. urens seed (Fig. 65 A-F) are like *M. sloanei* except for color. *M. urens* seeds are monochrome light reddish or grayish brown with a grayish border around the hilum. It is difficult to separate these seeds. Seeds of *M. urens* are much less frequent than those of *M. sloanei.* It is likely that most seeds labelled *M. urens* in the drift literature are *M. sloanei.*

M. gigantea seed (Fig. 65 G-K) 20 to 35 mm in diameter, 6 to 7 mm thick, round, strongly compressed in cross section, lustrous black, surface wrinkled. Hilum black, about 3 mm broad. Lightweight cotyledonary tissue buoys seeds for an undetermined amount of time. Whole seeds are used as watch charms, and powdered seeds are used as aphrodisiacs (Stafford, 1905).

M. nigricans seed (Fig. 65 L-N) 25 to 35 mm in diameter, about 20 mm thick, round, compressed in cross section, lustrous brown mottled with black, surface essentially smooth. Hilum brown, about 4 mm broad. Lightweight cotyledonary tissue and cavities at both ends buoy seeds for an undetermined amount of time. The seed coat is about twice as thick as the seed coats of the other species.

Other *Mucuna* spp. seeds are known to drift, including those shown in Figure 66. The species represented by A-D has an unusually broad hilum.

Leguminosae

MUCUNA SPP. I

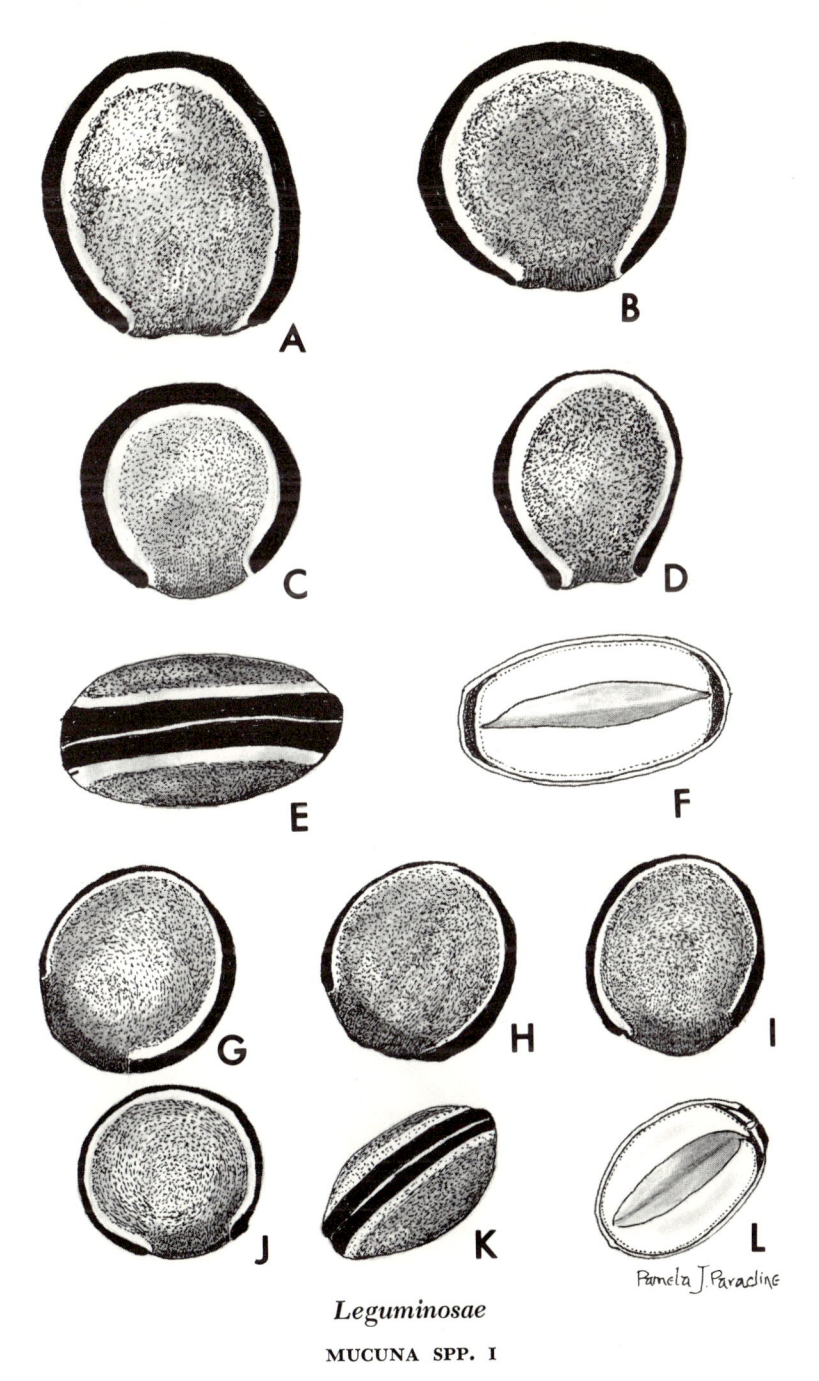

Figure 64. *Mucuna fawcettii* (A-F); *M. sloanei* (G-L). A-L, seeds. A-D, G-J, lateral views; E, K, hilar views; F, L, cross section (X1). Seeds, Gunn and Dennis, southeastern coast of Florida.

(159)

Leguminosae

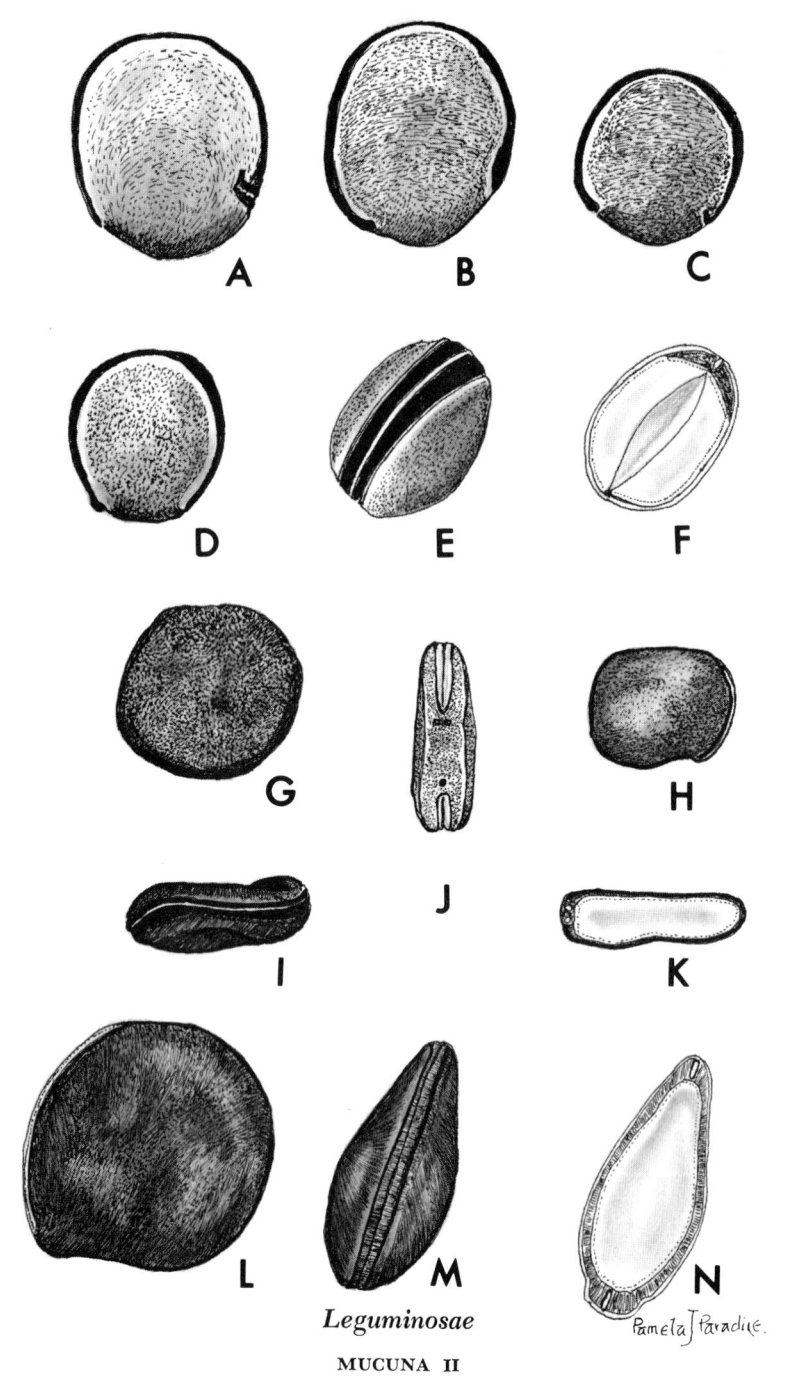

MUCUNA II

Figure 65. *Mucuna urens* (A-F); *M. gigantea* (G-K); *M. nigricans* (L-N). A-D, G, H, L, lateral views; E, I, M, hilar views; F, K, N, cross sections; J, micropyle view (X1). A-F, Gunn and Dennis, southeastern coast of Florida; G-N, Degener and Degener, beaches of Canton Island.

(160)

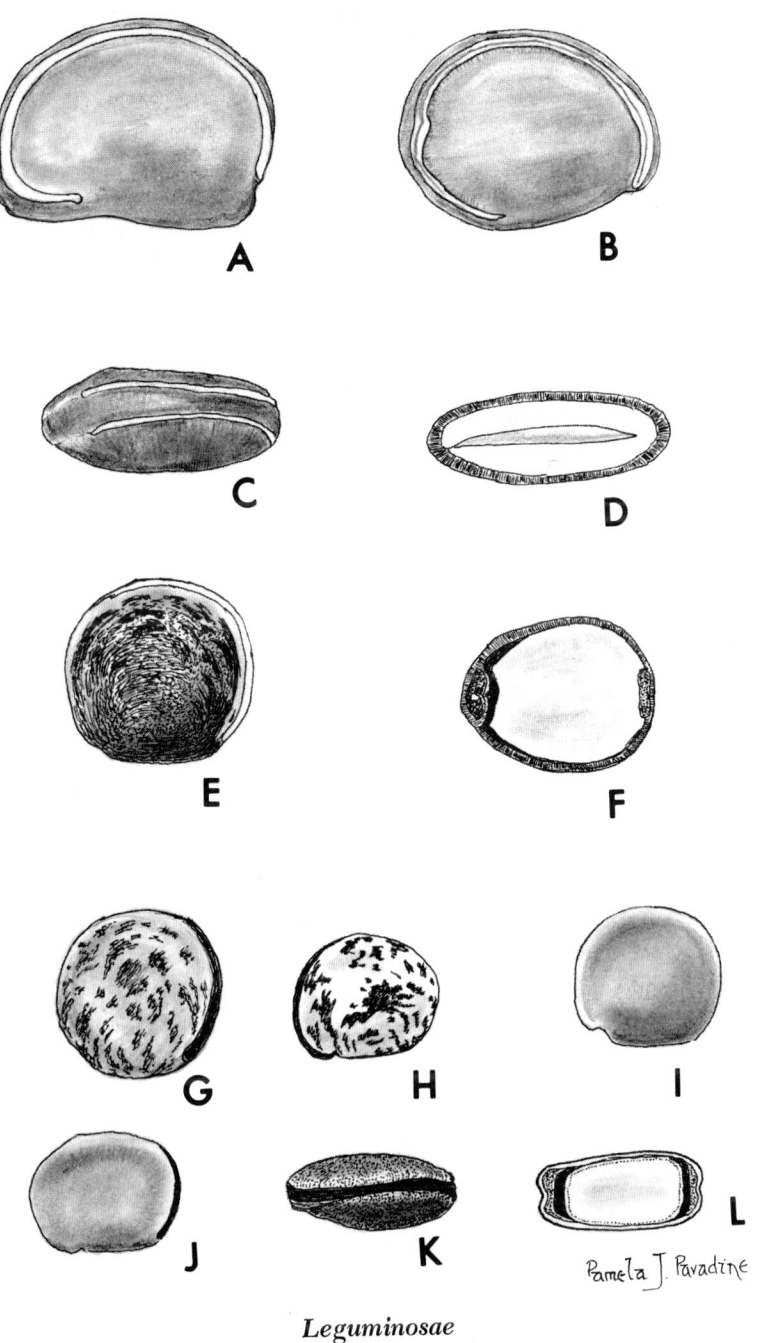

Leguminosae

MUCUNA III

Figure 66. *Mucuna* spp. seeds. A-B, E, G-J, lateral views; C, K, hilar views; D, F, L, cross section (X1). Seeds, Degener and Degener, beaches of Canton Island.

(161)

Leguminosae

OXYRHYNCHUS TRINERVIUS (DONN. SM.) RUDD
STRONGYLODON LUCIDUS (FORST. F.) SEEM.

Disseminule: Seed.
Description: *A. trinervius* seed (A-D) 1.5 to 2 cm in diameter, globose, round in cross section, lustrous black to dark brown, smooth. Hilum color of seed coat or lighter, with raised margins, and occupying about ¾ of the seed circumference. *S. lucidus* seed (E-H) 1.5 to 3 cm long, 1.5 to 2 cm wide, and 1 to 1.5 cm thick, round, compressed in cross section, lustrous black, smooth. Hilum color of seed coat or lighter, with flush to barely raised margins, and occupying about ⅔ of the seed circumference.
Buoyancy factor: Buoyant cotyledonary tissue.
Buoyancy: *O. trinervius*—at least 6 months; *S. lucidus*—not tested.
Viability: *O. trinervius*—most seeds viable; *S. lucidus*—not tested.
Currents: *O. trinervius*—tropical currents arising in the New World; *S. lucidus*—tropical currents arising in the Pacific region.

Oxyrhynchus trinervius seeds have been stranded along the coast of southeastern Florida and the Gulf of Mexico. These seeds are produced by an herbaceous to woody vine which is native to the area from southern Mexico to western Colombia (Rudd, 1967). Seeds of a related species, *O. volubilis* Brandegee of Texas, eastern Mexico, Cuba, and Bahamas are similar, though only up to 1 cm long. The hilum on these seeds is snow white. We do not know if these seeds float.

Strongylodon lucidus is a beautiful, high-climbing woody vine which is native to Polynesia, but which now has been distributed throughout the tropics. The vines are usually found along stream sides in lowland rain forests.

Leguminosae

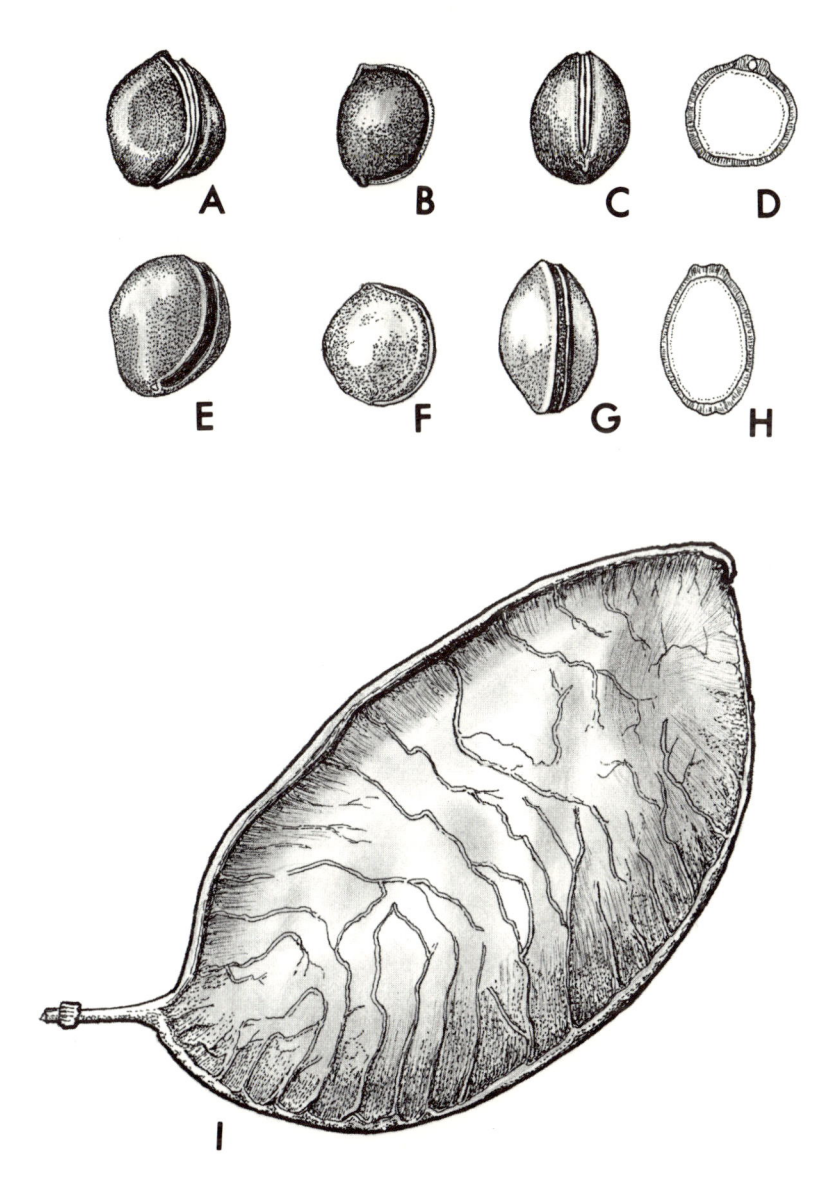

OXYRHYNCHUS SPP.

Figure 67. *Oxyrhynchus trinervius* (A-D); *Strongylodon lucidus* (E-I). A-H, seeds; I, fruit. A-B, E-F, I, lateral views; C, G, hilar views; D, H, cross sections (X1). A-D, Gunn and Dennis, southeastern coast of Florida; E-H, Degener and Degener, beaches of Canton Island; I, fruit, from vine.

Leguminosae

PELTOPHORUM INERME (ROXB.) NAVES, YELLOW FLAMBOYANT

Disseminule: Fruit.
Description: Fruit (A-F) 5 to 8 cm long, 1.5 to 2.5 cm wide, 3 to 4 mm thick, strongly compressed in cross section, reddish brown to straw colored, surface covered with minute parallel grooves, some of the central grooves may be cracked open.
Buoyancy factor: Corky central portion of fruit (E, F).
Buoyancy: About 9 months.
Viability: Most seeds viable.
Currents: Most tropical currents.

Yellow flamboyant tree, a native of the East Indies, has been widely planted throughout the tropics and subtropics for its colorful flowers and as a rapidly growing shade tree. The reddish brown surfaces of the mature fruits may be eroded in the stranded fruits, leaving a straw colored surface. Each indehiscent fruit contains 1 to 4 seeds which are freed only after the fruit decays. The seeds (G) do not float.

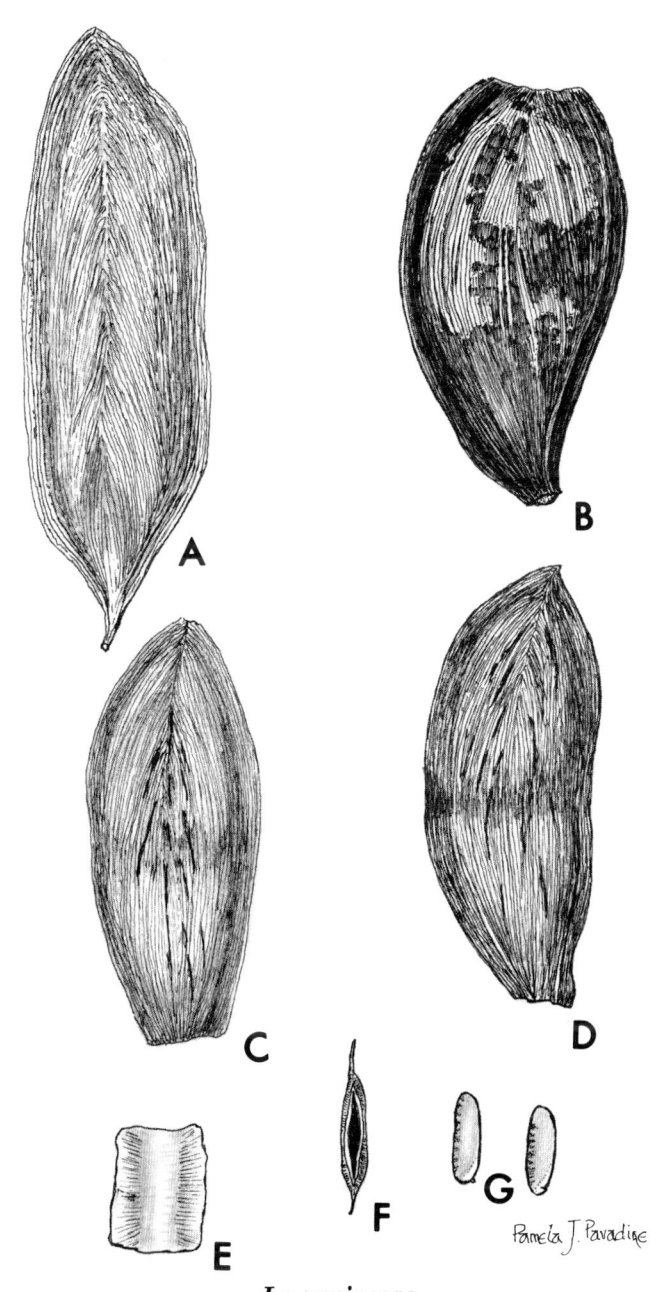

Leguminosae

PELTOPHORUM INERME (ROXB.) NAVES

Figure 68. *Peltophorum inerme* fruits and seeds. A-F, fruits; G, seeds. A-D, G, lateral views; E, internal "core" of fruit; F, cross section (X1). Fruits, Gunn and Dennis, southeastern coast of Florida (seeds extracted from drift fruits).

Leguminosae

PTEROCARPUS OFFICINALIS JACQ., BLOODWOOD

Disseminule: Fruit.
Description: Fruit (A-D) 3 to 8 cm long, 5 to 10 mm thick, irregularly rounded, strongly compressed in cross section, grayish brown, surface prominently veined due to erosion of smooth outer surface.
Buoyancy factor: Fibrous corky fruit coat and space around seed.
Buoyancy: About 1 year.
Viability: Few seeds viable.
Currents: Tropical currents arising in the New World and perhaps the Old World.

The common name, bloodwood, is derived from the red sap that exudes from cut wood. Bloodwood trees are native to the New World tropics. They have now been planted throughout the Old World tropics. As a member of the littoral flora, bloodwood trees often form colonies in swamps behind mangrove colonies. Fruits are produced in the winter, and by spring the swamp water below the trees may be covered with floating fruits. Some of these fruits reach the ocean and drift. Many of the fruits stranded on southeastern Florida beaches bear tubes of sea-worms, *Teredo* spp.

Occasionally small fruits (E-G) about 3.5 cm long and 2 cm wide are stranded on southeastern Florida beaches. These fruits resemble bloodwood fruits, and they may be smaller fruits of this species. On the other hand, they may be from another species. Neither fruit contributes much to the protection of the seed. The fruits are often full of holes and some are broken.

Leguminosae

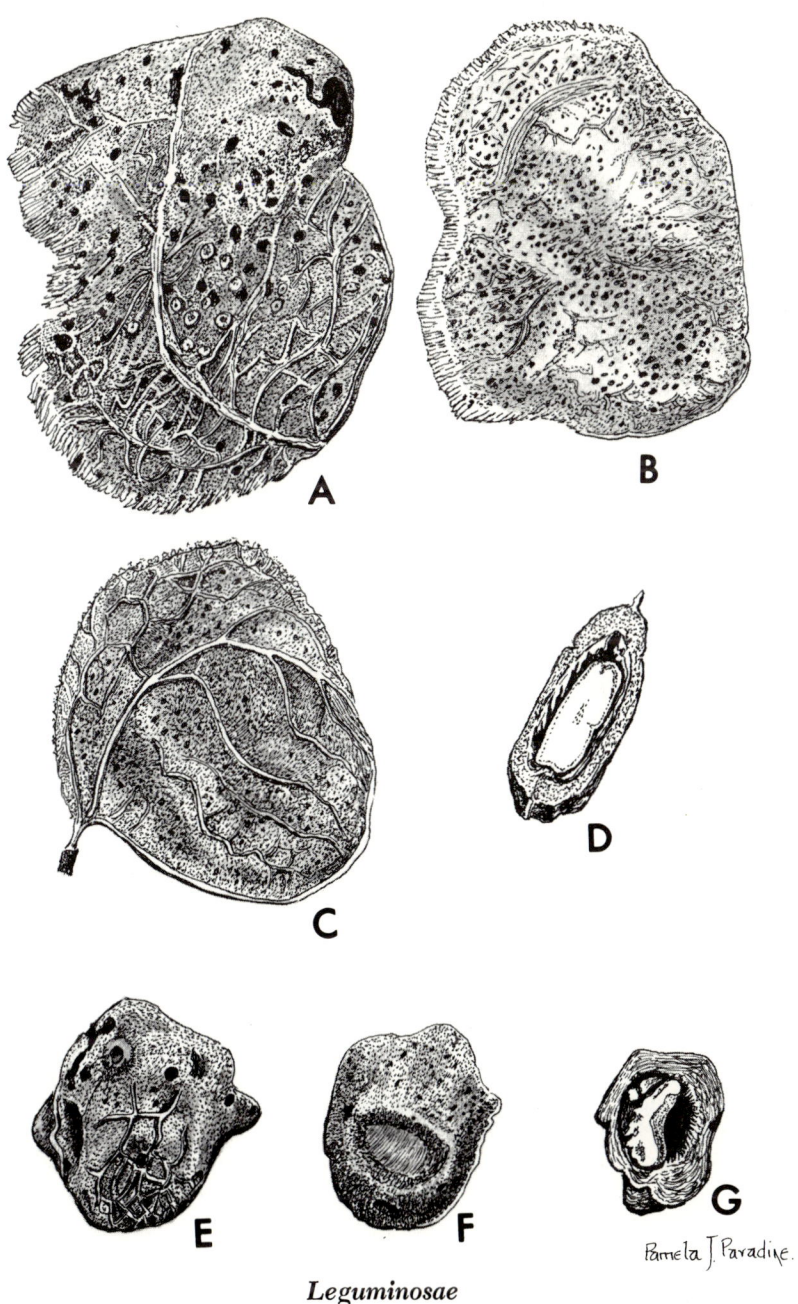

PTEROCARPUS SPP.

Figure 69. *Pterocarpus officinalis* (A-D); *P.* sp. ? (E-G). A-G, fruits. A-C, E-F, lateral views; D, G, cross sections (X1). Fruits, Gunn and Dennis, southeastern coast of Florida.

(167)

Meliaceae

CARAPA GUIANENSIS AUBLET, CRABWOOD
XYLOCARPUS MOLUCCENSIS (LAM.) ROEM., CANNONBALL TREE

Synonym: *Carapa* is *Xylocarpus* in some drift literature.
Disseminule: Seed.
Description: *C. guianensis* seed (A-C) 3 to 7 cm long, 3 to 6 cm wide, variously shaped (some triangular or compressed), outer margin (opposite hilum) curved, tan, brown, gray, or blackish, surface faintly roughened and bearing a faint and distorted to elongate hilum with a fibrous surface. *X. moluccensis* seed (E-F) like the preceding though usually slightly larger and often more compressed and hilum ellipsoidal or round and conspicuous.
Buoyancy factor: Embryo absent or nearly so.
Buoyancy: *C. guianensis*—more than 1.5 years; *X. moluccensis*—not tested.
Viability: *C. guianensis*—about 50 percent viable; *X moluccensis*—not tested.
Currents: *C. guianensis*—tropical currents arising in the New World; *X. moluccensis*—tropical currents arising in the southeast Asia and Pacific regions.

Crabwood is a native tree of the New World tropics, while cannonball tree (named for its round fruits) is a native of the Old World tropics. Both species are major members of the estuarine vegetation of their respective regions. Cannonball trees tend to favor mangrove swamps more than do the trees of crabwood. The outer portion of the seed coat of both species is composed of stone cells which give the seed coat durability during drifting. While local drifting may occur, long range drifting of live seeds apparently does not occur. This explains why *X. moluccensis* has not reached the New World tropics. The oil in the embryo (there is no endosperm) is used in industry and medicines, and as an illuminant.

A seed (D) similar to seeds of crabwood may be found stranded on New World beaches. This is the ivory nut (*Phytelephas macrocarpa* Ruiz & Pav.). While both seeds are about the same size, shape, and color, they may be separated by the hilum. The hilum of ivory nut is bold, round, and plate-like, while the hilum of crabwood is faint, often distorted, and fibrous.

Meliaceae

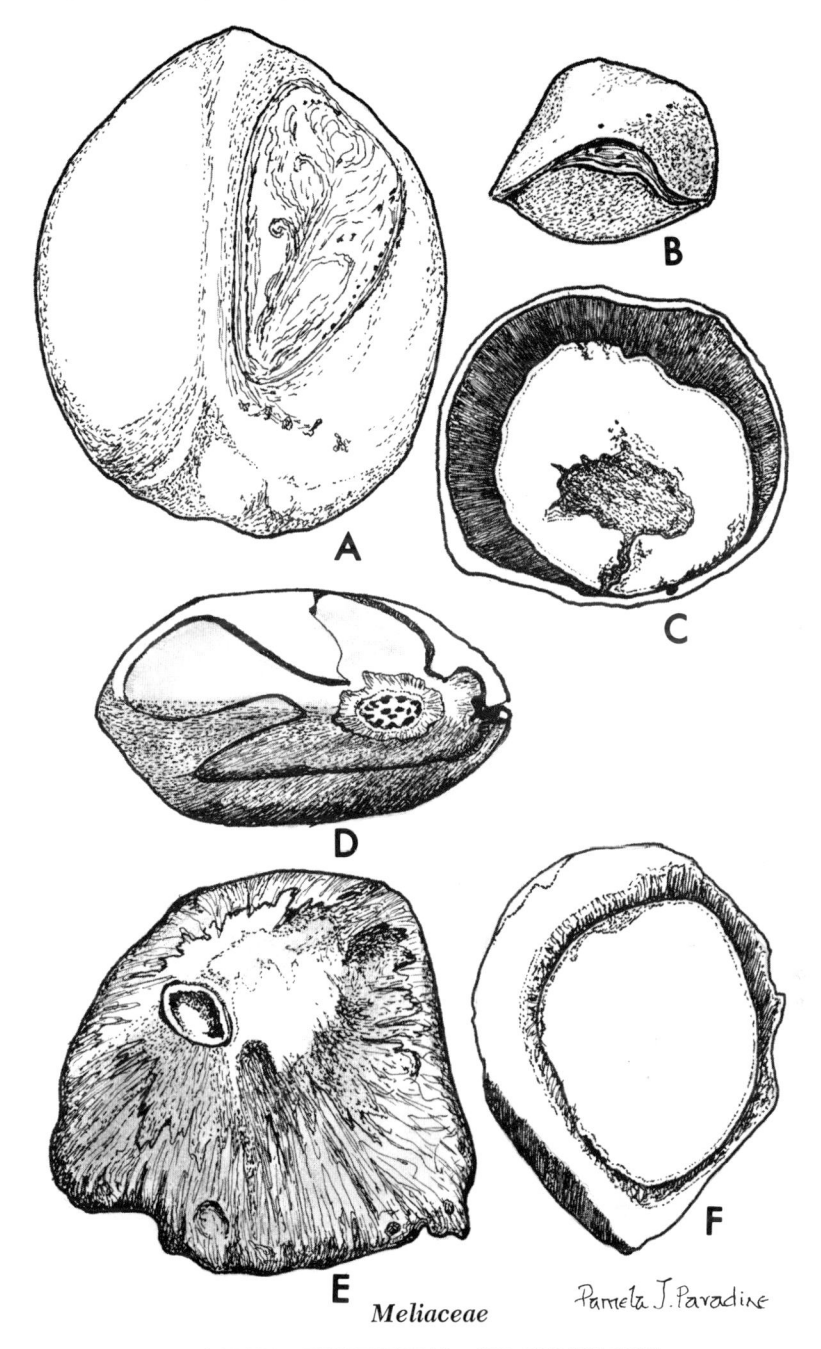

CARAPA, PHYTELEPHAS, AND XYLOCARPUS

Figure 70. *Carapa guianensis* (A-C); *Phytelephas macrocarpa* (D); *X. moluccensis* (E-F). A-F, seeds. A, B, D, E, hilar views; C, F, cross sections (X1). A-D, Gunn and Dennis, southeastern coast of Florida; E-F, Dennis, Jr., Singapore beach.

Meliaceae

SWIETENIA MAHAGONI (L.) JACQ., MAHOGANY

Disseminule: Fruit or individual carpel wall.
Description: Fruit (A-E) up to 10 cm long and 6 cm in diameter, ovate to pyriform, apex rounded, base bearing stem or stem scar, round in cross section, dark brown, surface smooth or 5 grooved, in cross section (E) revealing many winged seeds in 5 separate carpels.
Buoyancy factor: Corky fruit.
Buoyancy: Not tested.
Viability: Not tested.
Currents: Tropical currents arising in the New World.

Mahogany wood is a highly valued wood which takes a fine polish. The tree is also a valuable shade tree. A native of tropical America, mahogany trees have now been spread by man throughout the tropics of both Worlds. We believe that drifting has had little to do with this distribution, except in a limited way. The tree naturally occurs in moist to dry habitats of the interior lowlands. The fruits are carried by rivers to the ocean where they may drift. Guppy (1917) reported that freed seeds do not drift more than two weeks before dying. We have found no stranded seeds. The entire fruits may be solid or heavily pitted (B). Individual sections of the fruit (D) are stranded, indicating that the fruit wall is buoyant.

Meliaceae

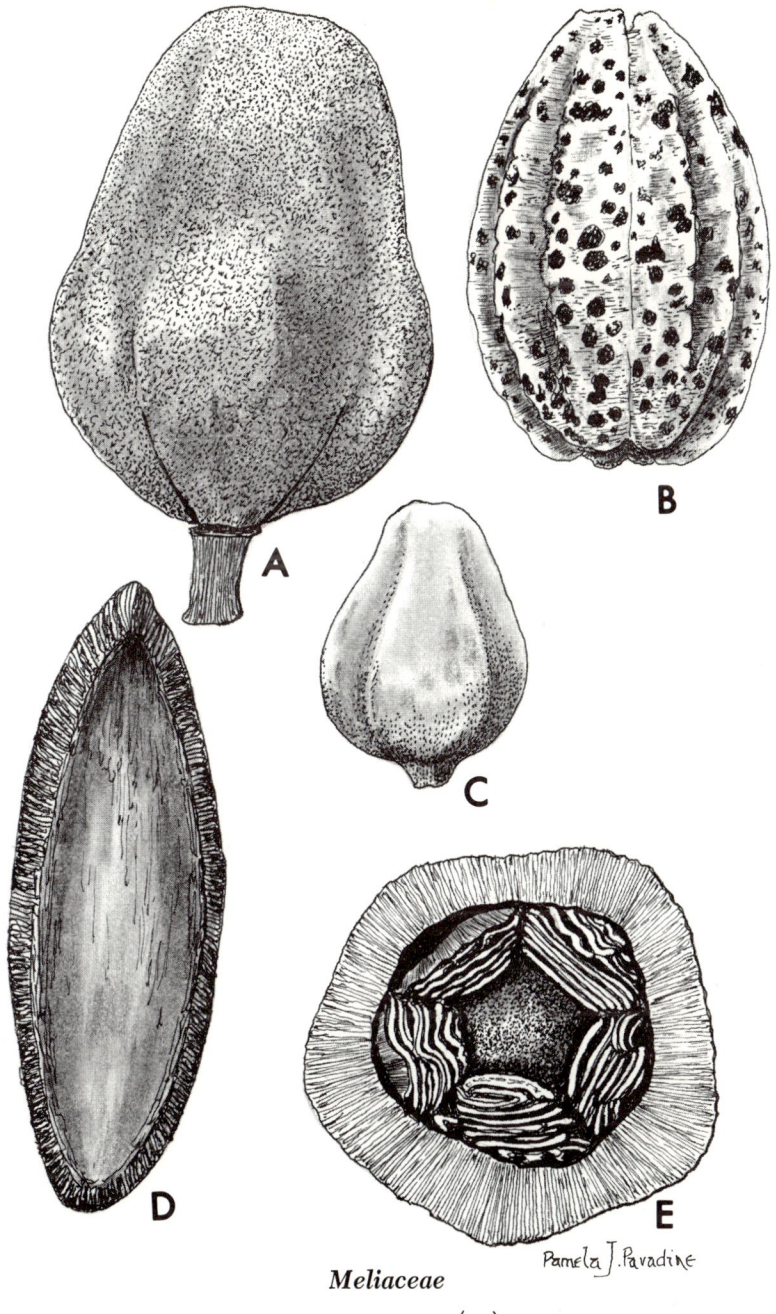

SWIETENIA MAHAGONI (L.) JACQ.

Figure 71. *Swietenia mahagoni.* A-C, E, fruits; D, carpel wall. A-D, lateral views; E, cross section (X1). Fruits and carpel wall, Gunn and Dennis, southeastern coast of Florida.

(171)

Palmae

ACROCOMIA SPP., PRICKLY PALMS

Disseminule: Endocarp.
Description: Endocarp (B-I) 2 to 3.5 cm in diameter, globose to subglobose, round in cross section, black, gray, or tan, surface smooth to slightly pitted, bearing 3 equidistant pores (holes) along its equator.
Buoyancy factor: Empty endocarp.
Buoyancy: At least 2 years.
Viability: Seed absent or nearly so.
Currents: Tropical currents arising in the New World.

Prickly palms, also known as corozo palms, are native trees of tropical America and the West Indies. While isolated endocarps cannot be identified to species, they may be easily separated from other palm endocarps by the 3 equitorial pores. These pores, which do not open into the seed cavity, may be prominent (D-H) or inconspicuous. Fresh fruits bearing seeds are too heavy to float. Buoyancy is acquired as the seed decays within the endocarp, and the exocarp (A) erodes. Buoyant endocarps are often empty and many may have been formed without a seed. When the seed is present, its edible oily endosperm resembles and tastes like coconut (*Cocos nucifera*) endosperm.

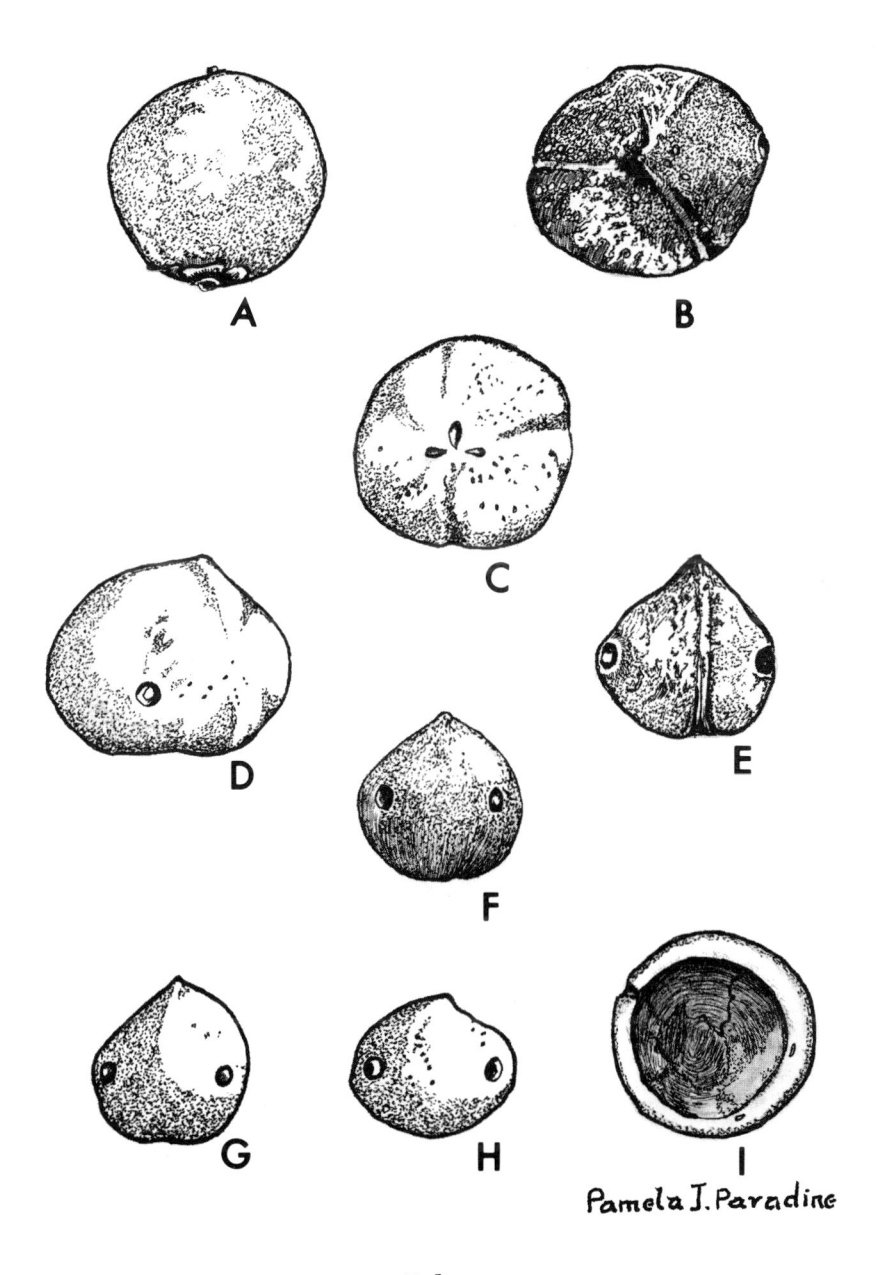

Palmae

ACROCOMIA SPP.

Figure 72. *Acrocomia* spp. A, fruit; B-I, endocarps. A, D-H, lateral views; B, apical view; C, basal view; I, cross section (X1). Fruit, from tree; endocarps, Gunn and Dennis, southeastern coast of Florida.

(173)

Palmae

ASTROCARYUM SPP., STARNUT PALM

Disseminule: Endocarp.
Description: Endocarp (B-H) 2 to 6 cm long, 2 to 4 cm in diameter at apex, obovoid, pointed at apex, rounded at base, round in cross section, black to brownish black, surface bearing numerous parallel and often light colored striations which radiate from the basal pore area and occasionally may be worn smooth, and 0 to 3 basal pores (F-G).
Buoyancy factor: Empty endocarp.
Buoyancy: About 22 months.
Viability: Seed absent or nearly so.
Currents: Tropical currents arising in the New World and perhaps some tropical currents in the Pacific region.

Isolated endocarps of starnut palms, or black palms, cannot be identified to species, though we suspect that most endocarps on New World beaches are from trees of *A. mexicana* Liebm. and *A. standleyana* L. H. Bailey. Some species of *Astrocaryum* are native to tropical America and to the West Indies where they are often members of the wet lowland forest. While viable endocarps and fruits may be transported by rivers, ocean drifting does not play a role in a starnut palm spread, because the endocarps are consistently empty. The bony endocarps may be polished and fashioned into jewelry. The endocarp is covered by a thin fibrous exocarp which usually bears a prominent apical beak and an indurate basal calyx (A). Starnut palm endocarps resemble endocarps of the African oil palm, *Elaeis guineensis* Jacq. Some of the drift endocarps which lack the basal striations may be endocarps of the African oil palm.

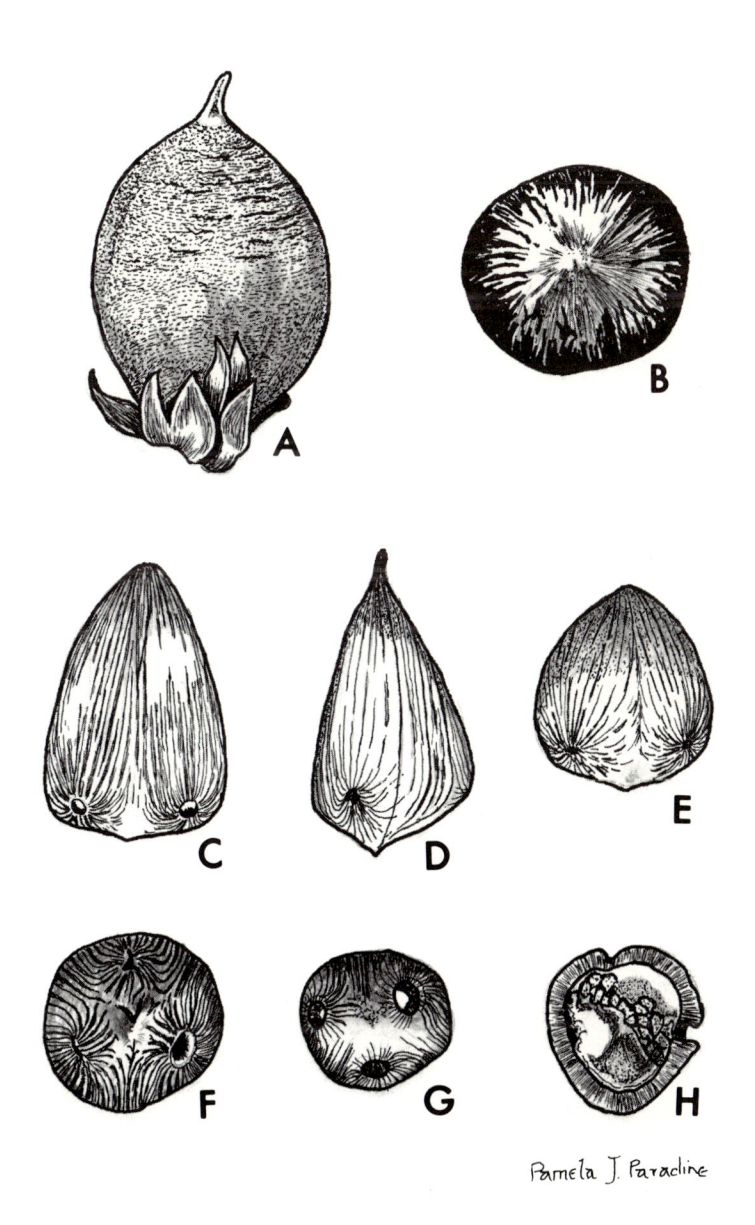

Palmae

ASTROCARYUM SPP.

Figure 73. *Astrocaryum* spp. A, fruit; B-H, endocarps. A, C-E, lateral views; B, apical view; F-G, basal views; H, cross section (X1). Fruit, from tree; endocarps, Gunn and Dennis, southeastern coast of Florida.

(175)

Palmae

BORASSUS SPP.

Disseminule: Endocarp.

Description: Endocarp (A-C) up to 13 cm long, up to 6 cm in diameter, oblong, apical scar large and irregular, base deeply notched, notch bearing (A), or devoid (B) of, blackish brown fibers, compressed in cross section, blackish brown, surface nearly smooth.

Buoyancy factor: Empty endocarp.
Buoyancy: Not tested.
Viability: No seed present.
Currents: Tropical currents arising in the southeast Asia and Pacific regions.

While Harold Moore, Jr., Bailey Hortorium, Cornell University, confirmed the generic identification, he was unable to assign the endocarps to one or more species. He suspects that more than one species is represented in the collection. *Borassus* spp. are large trees, native of Africa and Asia. One species, *B. flabelliformis* L., is widely cultivated in the Indian Ocean region. It is surprising that Muir (1937) did not find stranded endocarps at his South African stations.

Palmae

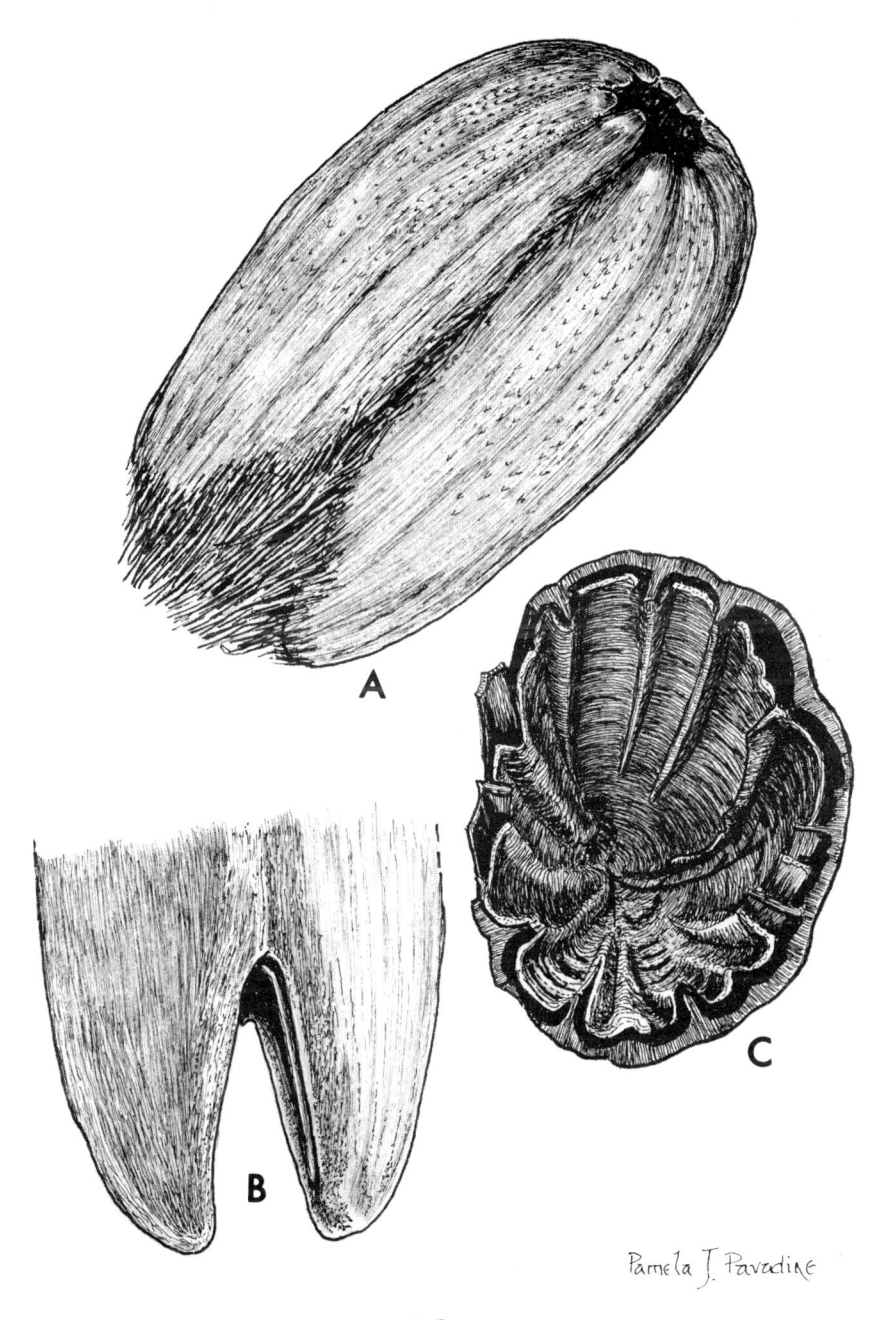

BORASSUS SPP.

Figure 74. *Borassus* spp. endocarps. A, lateral (notch bearing fibers); B, partial lateral view (notch fiberless); C, cross section (X1). Endocarps, Degener and Degener, beaches of Canton Island.

Palmae

COCOID PALM SPECIES
MAXIMILIANA CARIBAEA GRISEB. & WENDL.
ORBIGNYA COHUNE (MARTIUS) STANDLEY, COHUNE

☙

Disseminule: Endocarp.
Description: Endocarps often bearing 3, rarely 0 or 2 basal pores. Additional data presented below.
Buoyancy factor: Empty or partially filled endocarps; occasionally corky endocarp wall.
Buoyancy: Often one year or more.
Viability: Seeds seldom present. If present, usually not viable.
Currents: Tropical currents arising in the New World and perhaps other currents.

Cocoid palms are members of the subfamily Cocosoideae. This subfamily is composed of 27 genera and some 600 species: most native to the New World tropics, a few to Africa, and the coconut (*Cocos nucifera*) to the Indo-Malaysian region. In addition to the coconut, two other cocoid palms are discussed separately, *Acrocomia* spp. and *Astrocaryum* spp. The cocoid palm endocarps shown in Figure 75 were identified by Harold Moore, Jr., Bailey Hortorium, Cornell University. These stranded endocarps are either woody or stonelike and bear 3 (rarely 0 to 2) basal pores. They range in color from black to tan and usually have a smooth surface.

Maximiliana caribaea endocarp (Fig. 76 A-F) 3 to 5 cm long, 2 to 3 cm in diameter, ellipsoidal to obovoid, apex long tapered, base short tapered and bearing 3 (rarely 1 to 2) inconspicuous subterminal pores, round in cross section, tan to blackish, surface smooth. This is a native tree of Trinidad and adjacent islands. Even though the endocarps drift to Florida, the species has not spread by drifting, because seeds are absent or dead. Endocarps are often damaged when stranded (A-C).

Orbignya cohune endocarp (Fig. 77 B-F) 3 to 8 cm in length and diameter, obovate, ellipsoidal to nearly circular, round in cross section, tan to brown, surface striate and usually bearing 3 pores (E). These trees, native of the Atlantic coast lowland forest from southern Mexico to Central America, vary from almost trunkless to 15 m tall. A fruiting panicle may contain 800 to 1000 fruits and weigh up to 45 kg. The leaves may be 10 to 18 m long and 2 m wide. They are thought to be the largest leaf of any American spermatophyte.

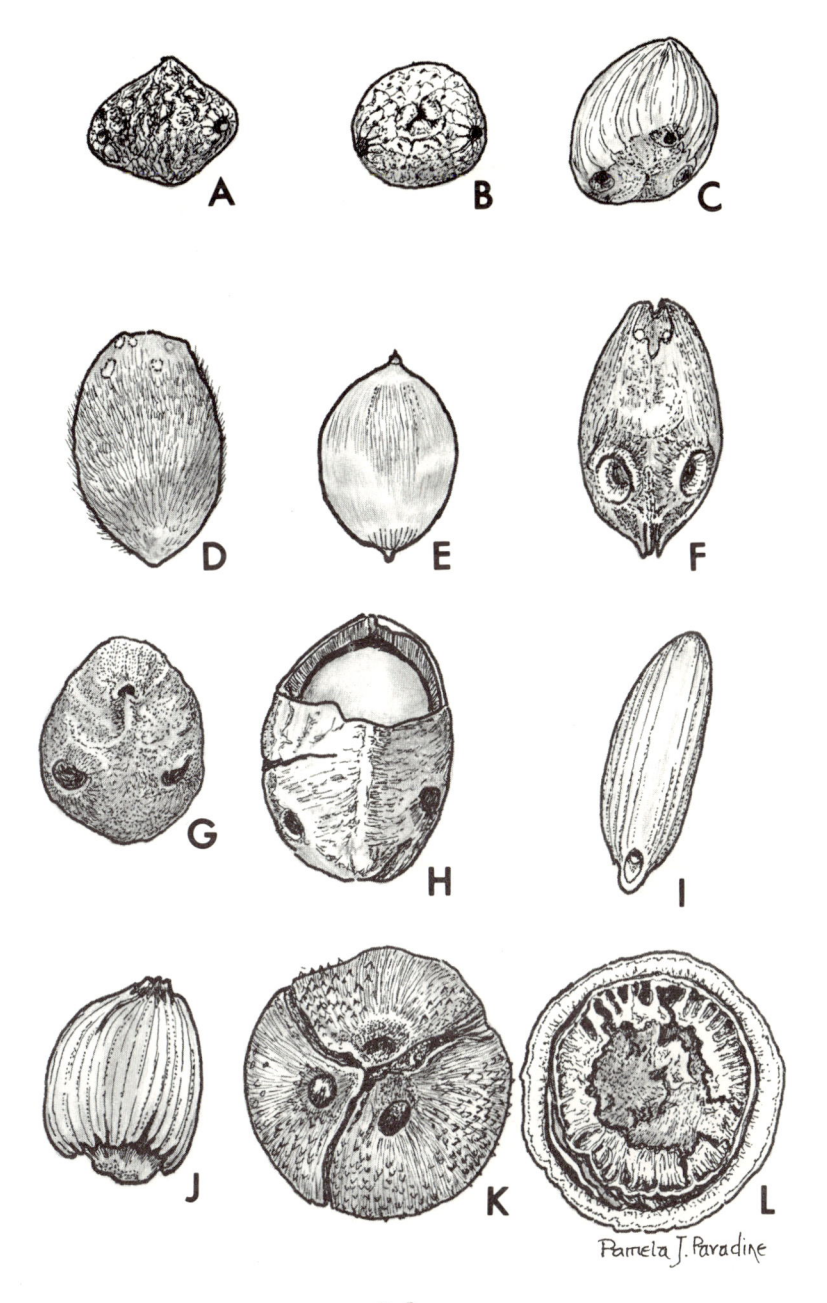

Palmae

COCOID PALMS

Figure 75. Cocoid palm spp. endocarps. A-H, lateral views; K, basal view; L, cross section (X1). Endocarps, Gunn and Dennis, southeastern coast of Florida.

(179)

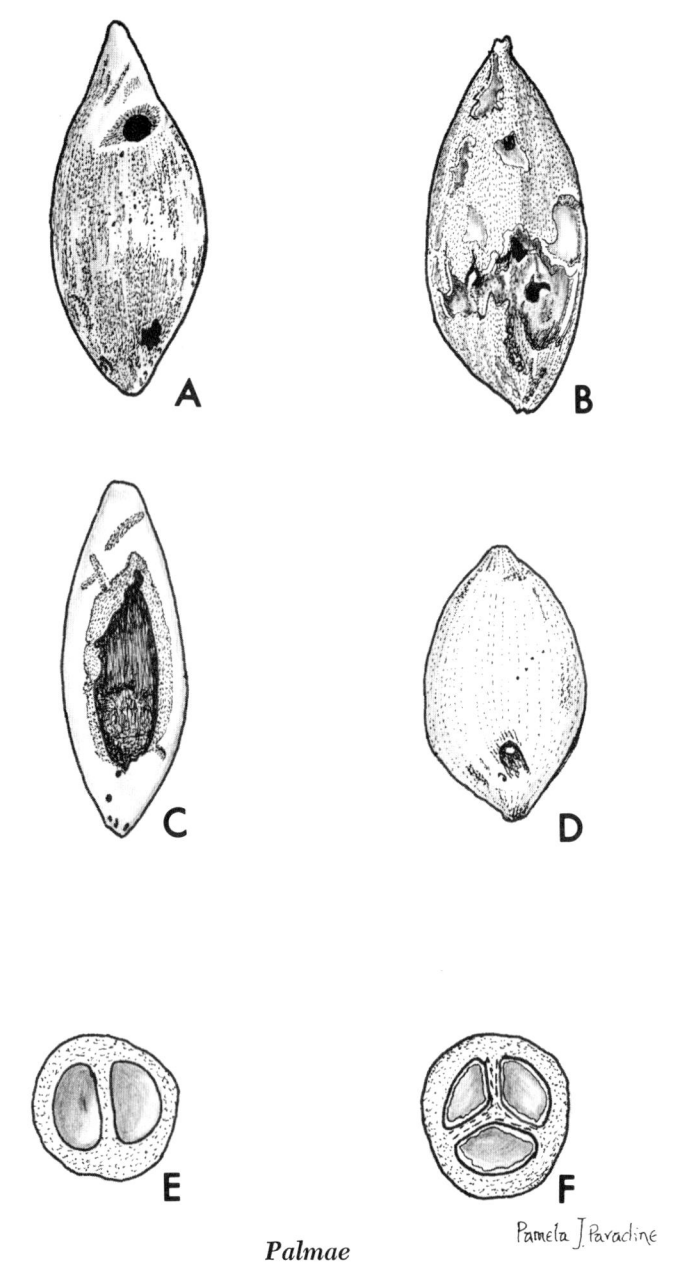

Palmae

MAXIMILIANA CARIBAEA GRISEB. & WENDL.

Figure 76. *Maximiliana caribaea* endocarps. A-D, lateral views; E-F, cross sections showing 2 and 3 seed cavities (X1). Endocarps, Gunn and Dennis, southeastern coast of Florida.

Palmae

ORBIGNYA COHUNE (MARTIUS) STANDLEY

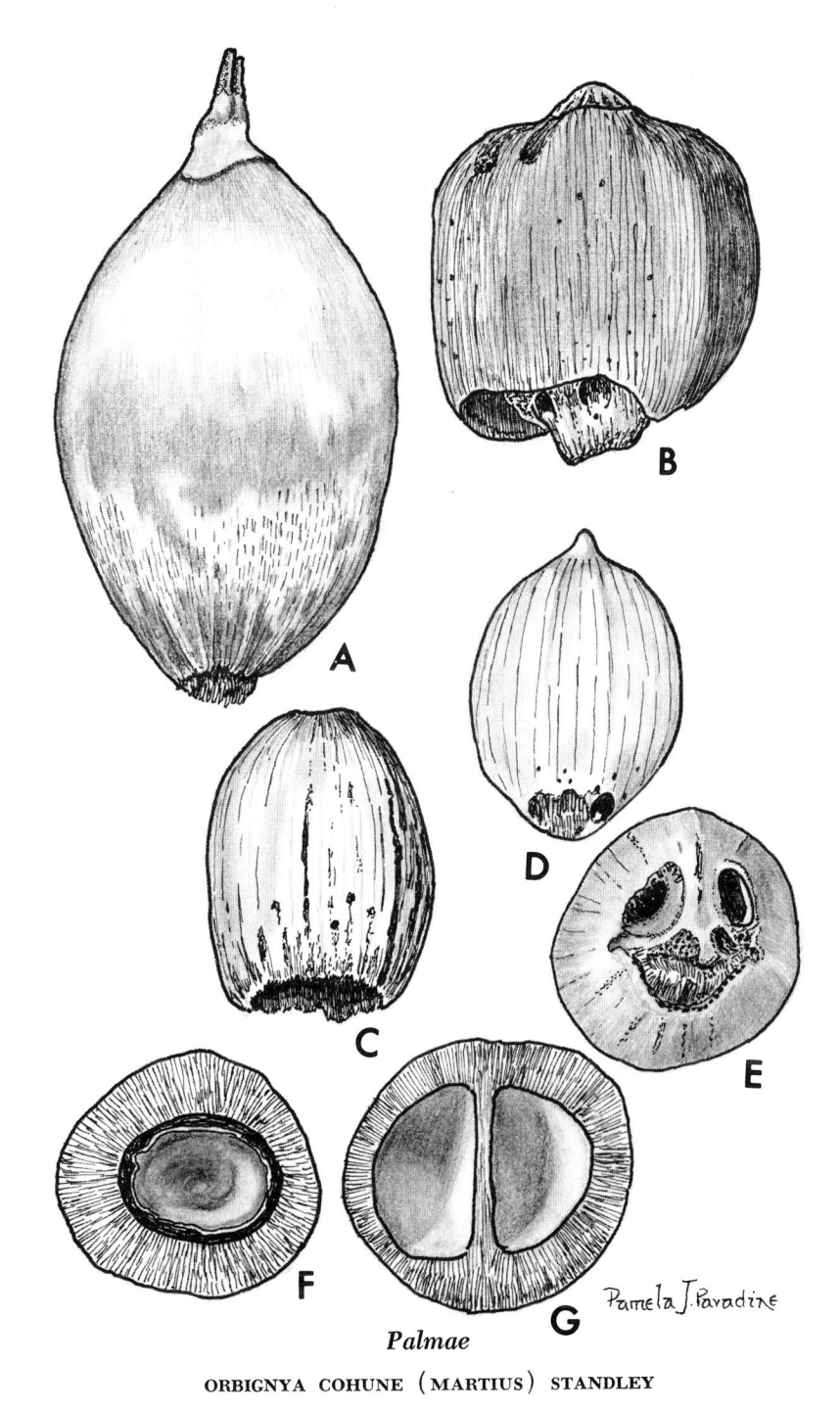

Figure 77. *Orbignya cohune* A, fruit; B-G, endocarps. A-D, lateral views; E, basal view showing 3 large pores; F-G, cross sections showing 1 and 2 seed cavities (X1). Fruit, from tree; endocarps, Gunn and Dennis, southeastern coast of Florida.

(181)

Palmae

COCOS NUCIFERA L., COCONUT

Disseminule: Mature or immature fruit.

Description: Mature fruit (Figs. 78 and 79) 10 to 40 cm long, 10 to 16 cm in diameter, globose to ellipsoidal, tapering to a nearly triangular apex, tapering to a truncate base bearing a prominent terminal scar, round to triangular in cross section, yellowish brown to dark brown, surface relatively smooth to more or less fibrous. Fruit and seed layers are labelled in Figure 78. Fruit layers: exocarp, a smooth thin outer layer; mesocarp, a fibrous buoyant layer at least 2 cm thick; endocarp, a bony, dark brown layer bearing 3 basal pores and a 3-parted raphe (B). Seed layers: a seed coat thin, brown; an endosperm, white and solid, liquid in young coconuts. Immature fruits (Fig. 80) may be miniatures of the mature fruits or endocarps (C-E), or be juvenile fruits still encased in the flower scales (A-B).

Buoyancy factor: Mature fruit—fibrous mesocarp and seed cavity; immature fruit—seed cavity.

Buoyancy: Mature fruit—with or without husk, at least 2 years; immature fruit—not tested.

Viability: Not tested. Germinating stranded and drifting coconuts have been observed.

Currents: Most tropical currents.

The coconut fruit is the best known drift fruit, though perhaps not the long-distance drifter some have thought it to be (Dennis and Gunn, 1971). Like *Cassia fistula* fruits, coconuts were found on Norwegian beaches and thought to have been carried there by the Gulf Stream. The discussion and citations given under *C. fistula* apply here. We do not doubt that coconuts, especially some races in the Pacific region, may be able to drift for thousands of miles. These Pacific races are more seaworthy than other races, because the mesocarp is thicker. The coconut is a ubiquitous tree of the tropics and subtropics, because of man's efforts and to a less extent drifting. Ranked as one of the ten most important tree species, the coconut contributes wood, thatching, coir, copra, food, drink, oil, and endocarps which are fashioned into decorative and useful articles.

Palmae

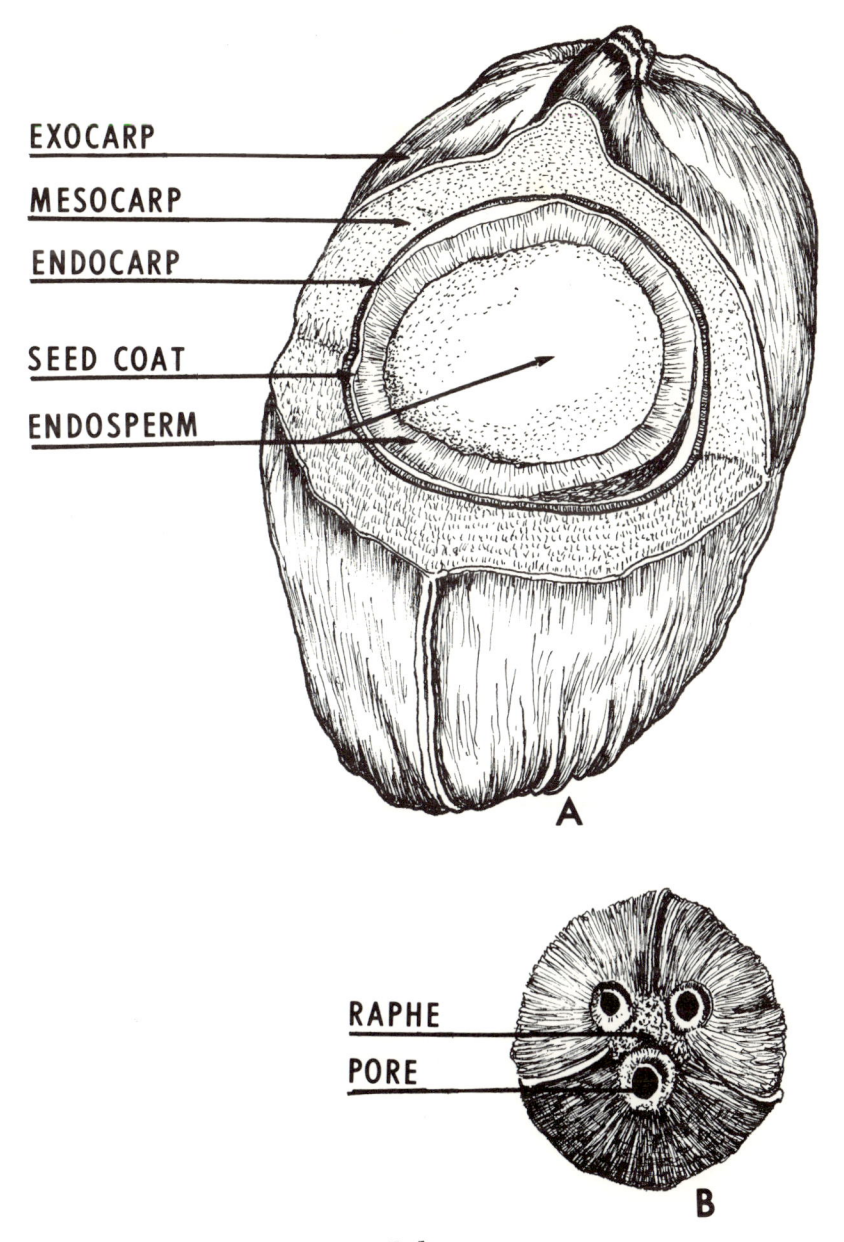

COCOS NUCIFERA L. I

Figure 78. *Cocos nucifera.* A, fruit, B, endocarp. A, lateral view; B, basal view (X½). Mossman, southeastern coast of Florida.

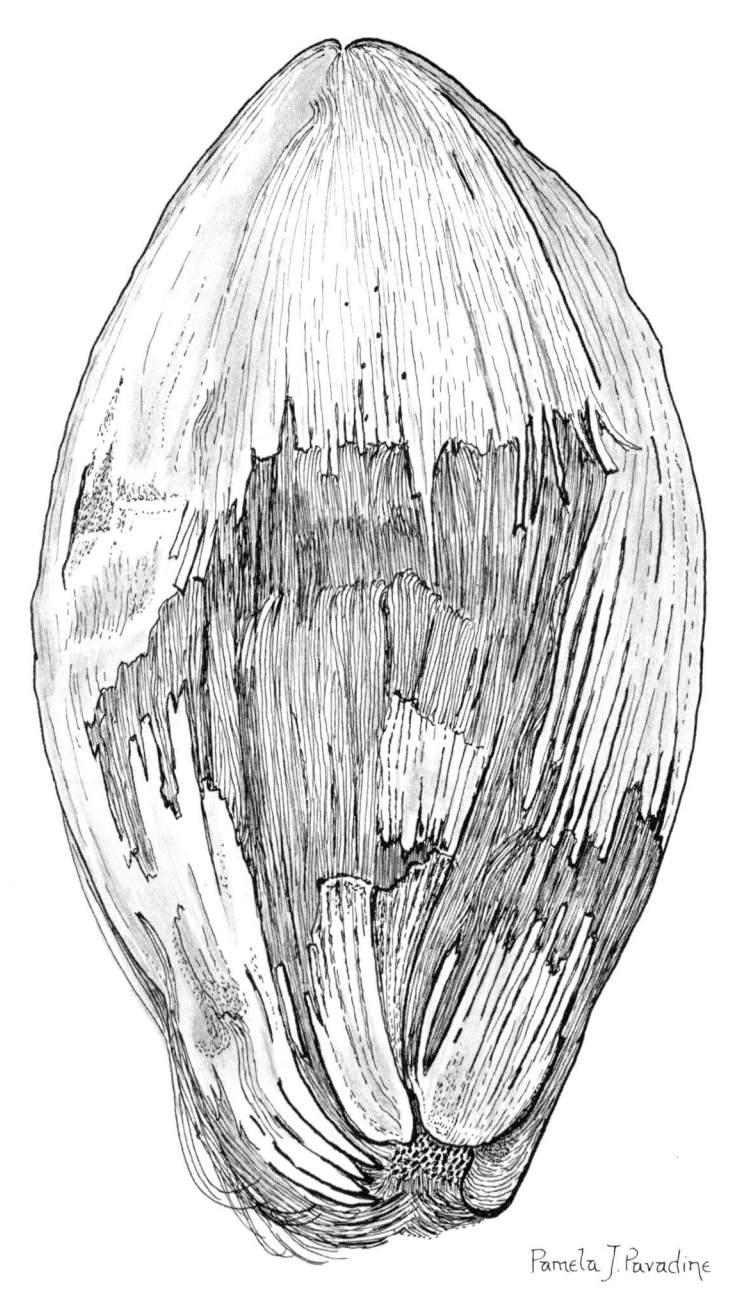

Palmae

COCOS NUCIFERA L. II

Figure 79. *Cocos nucifera*. Mature fruit (X½). Mossman, southeastern coast of Florida.

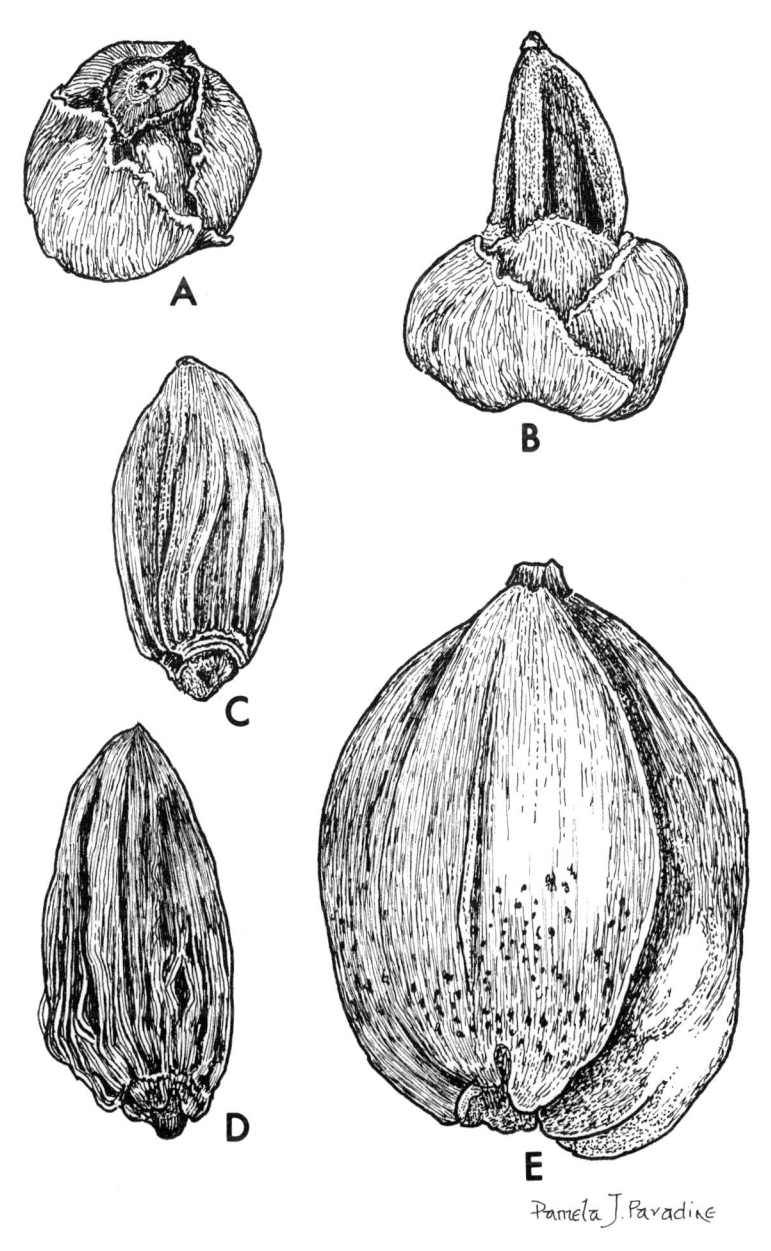

Palmae

COCOS NUCIFERA L. III

Figure 80. *Cocos nucifera*, immature fruits. A-B, fruits partially enclosed by bud scales; C-E, juvenile fruits. A, apical view; B-E, lateral views (X1). Mossman, southeastern coast of Florida.

(185)

Palmae

LODOICEA MALDIVICA (GMELIN) PERSOON, COCO-DE-MER

Synonym: *L. seychellarum* Labill.
Disseminule: Endocarp.
Description: Endocarp (Figs. 81–83) up to 30 cm long, up to 80 cm in circumference, frequently with a 2-lobed apex and base (rarely unlobed or 3-lobed), slightly compressed in cross section (rarely triangular), brown to blackish brown, surface irregularly and shallowly furrowed.
Buoyancy factor: Empty space within seed. Fresh endocarps, like the half shown in Figure 7, do not float, because their specific gravity is about 1.2. As the white endosperm is consumed by the ever growing embryo, or decays, the weight and the specific gravity decrease until the endocarp becomes buoyant.
Buoyancy: Not tested.
Viability: Not tested. No stranded endocarp is known to have a viable seed.
Currents: Currents passing the islands of Praslin and Curieuse in the Seychelles archipelago carry endocarps to the Maldive Islands and the Indian peninsula.

Coco-de-mer is also known as the sea-coconut and double-coconut, though it is not a close relative to the true coconut (*Cocos nucifera*). The coco-de-mer is closely related to *Borassus* and *Manicaria*. Drift endocarps were known for hundreds of years before the parent trees were discovered on adjacent islands in the Seychelles archipelago. The archipelago is about 1000 km northeast of Madagascar. One of the first published records of this giant endocarp was made by Pyrard (1611) who described endocarps collected from the beaches of the Maldive Islands. These islands lie 2300 km northeast of the Seychelles and about 800 km southwest of India. During the sixteenth and seventeenth centuries, it was generally believed that the endocarps possessed almost magical qualities as a medicine and as an aphrodisiac. Until the Seychelles were discovered by Mahé de la Bourdonnais in 1743, the endocarps were sold for exorbitant prices. Even after the islands were discovered, individual endocarps were selling for about 400 pounds sterling (Rochon, 1769). While the 1973 price of natural endocarps is only a few dollars, they remain a source of income for the Seychellois. The endocarps are often polished, hinged, and decorated and sold for a much higher price (Fig. 7). The gelatin-like, bland inner endosperm is served as a delicacy by the Seychellois.

Palmae

LODOICEA MALDIVICA (GMELIN) PERSOON I

Figure 81. *Lodoicea maldivica* endocarp with protruding embryonic root (X⅜). Endocarp, Lemon, Praslin Island.

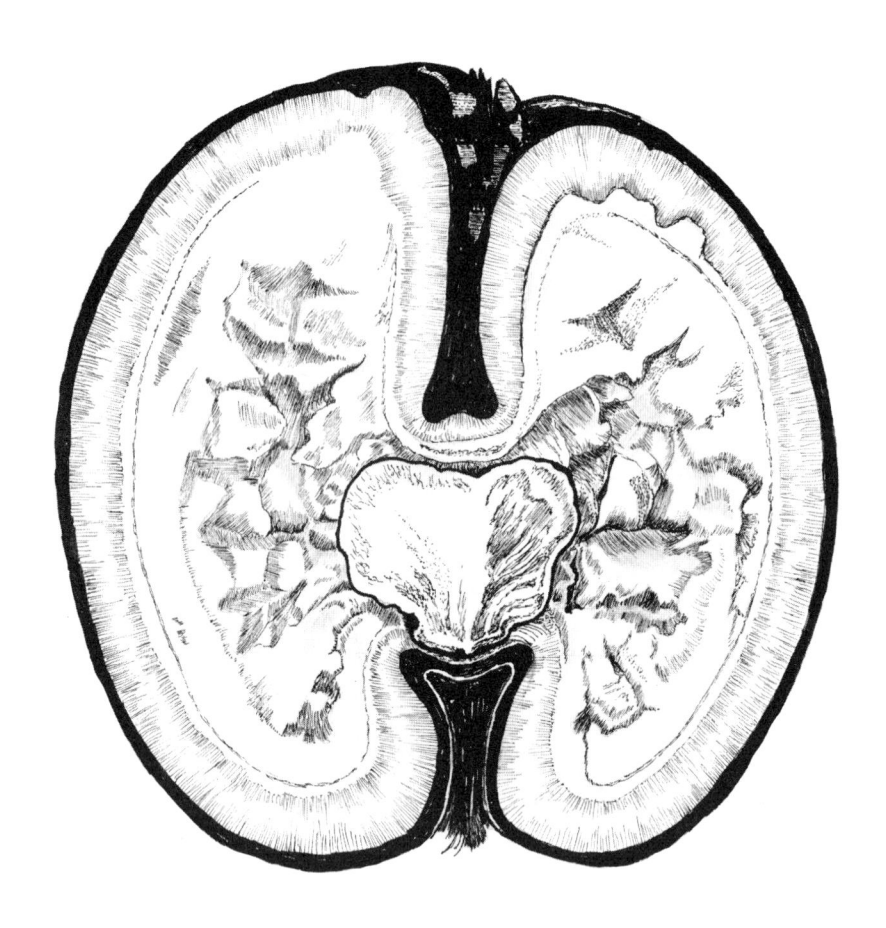

Palmae

LODOICEA MALDIVICA (GMELIN) PERSOON II

Figure 82. Fresh *Lodoicea maldivica* endocarp cut longitudinally to show the two layers of white endosperm, an inner soft "broken" layer and an outer bone hard layer. The haustorium, between the infolded endocarp, absorbs the endosperm as the embryo enlarges (X⅜). Endocarp, Lemon, Praslin Island.

Palmae

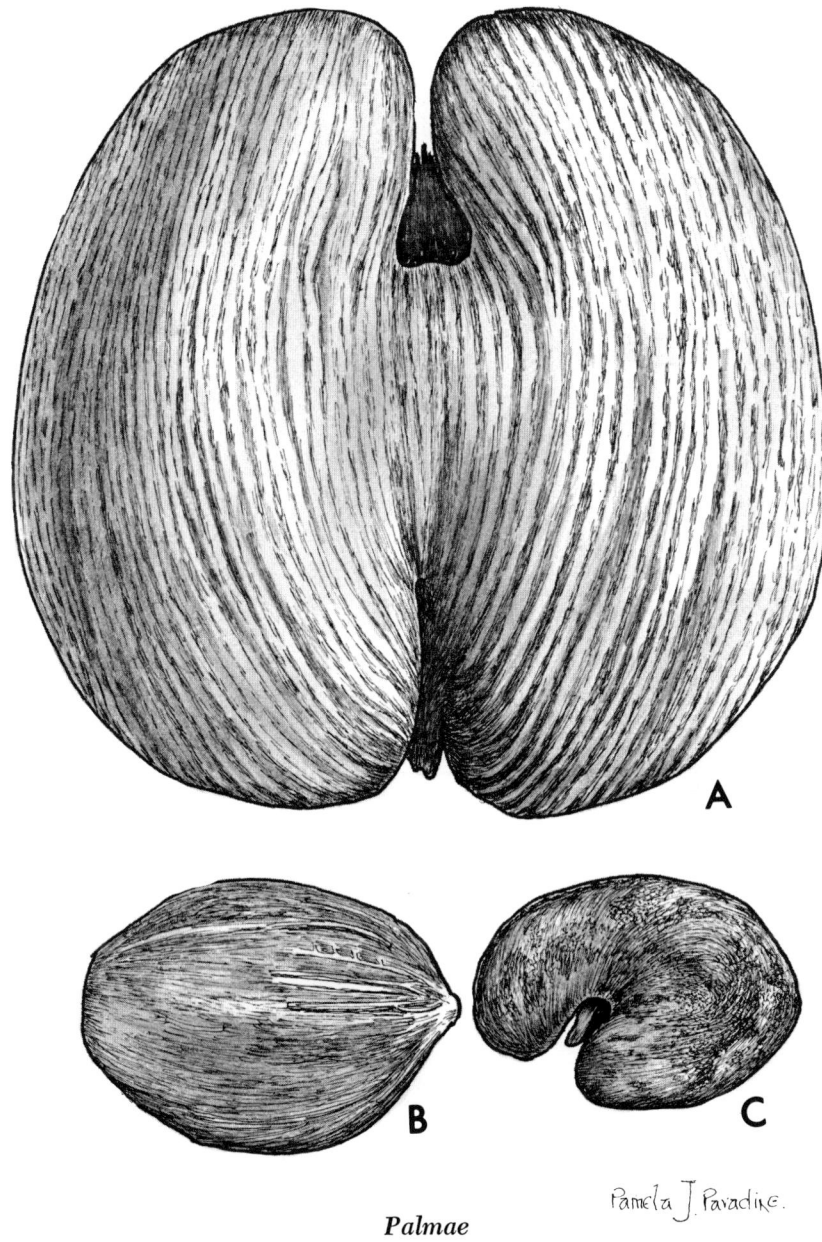

LODOICEA MALDIVICA (GMELIN) PERSOON III

Figure 83. The 3 largest disseminules: A, *Lodoicea maldivica;* B, *Cocos nucifera;* and C, *Mora oleifera* (X⅜). Disseminules, from trees.

(189)

Palmae

MANICARIA SACCIFERA GAERTNER, SEA-COCONUT

❦

Synonym: Bailey (1947) suggested that *M. plukenetii* G. & W. is a separate species which has disseminules similar to *M. saccifera*.

Disseminule: Endocarp or fruit.

Description: Fruit (A, D) 2 to 8 cm in diameter or length, 1-seeded is globose, 2-seeded is dumbbell-shaped, and 3-seeded is 3-lobed (A), round in cross section, gray-brown, surface heavily tuberculated. Endocarp (B, C, E-F) 1.5 to 5 cm in diameter, globose, round in cross section, brown, gray, blackish, white, or any combination of these colors, surface smooth to eroded and thus rough, usually bearing a conspicuous yellowish or light colored scar (B). Surface and color variation are due to the presence or absence of various fruit and seed layers.

Buoyancy factor: Cavity within endosperm.

Buoyancy: At least 2 years.

Viability: Most seeds viable.

Currents: Tropical currents arising in the New World and perhaps in the Old World.

The sea-coconut is a near relative of the coco-de-mer (*Lodoicea maldivica*) but not of the coconut (*Cocos nucifera*). The sea-coconut, also known as sleeve palm, is a tall tree of the American tropics. The species has been spread by man to the Old World tropics. No transport current is available to carry these disseminules to the Old World tropics. The disseminules have reached northern European beaches via the Gulf Stream (Sloane, 1707) where the climate is lethal to seedlings and young plants. Guppy (1917) reported the embryo to be short-lived. We can neither confirm nor reject this statement. The disseminules are quite variable, depending on which layer of the fruit or seed is exposed. This variation coupled with the variable number of seeds (1 to 3) per fruit makes these disseminules difficult to identify. Gray (almost white) stranded endocarps have an appropriate common name, golf balls. The entire fruit with its unusual appearance led Jonston (1662) to believe that it was an insect gall. The sea-coconut tree is unique, because it possesses the largest entire leaf blade of any plant (Corner, 1966). While usually growing in colonies near the seacoast, this tree makes an interesting ornamental when planted as a specimen tree.

Palmae

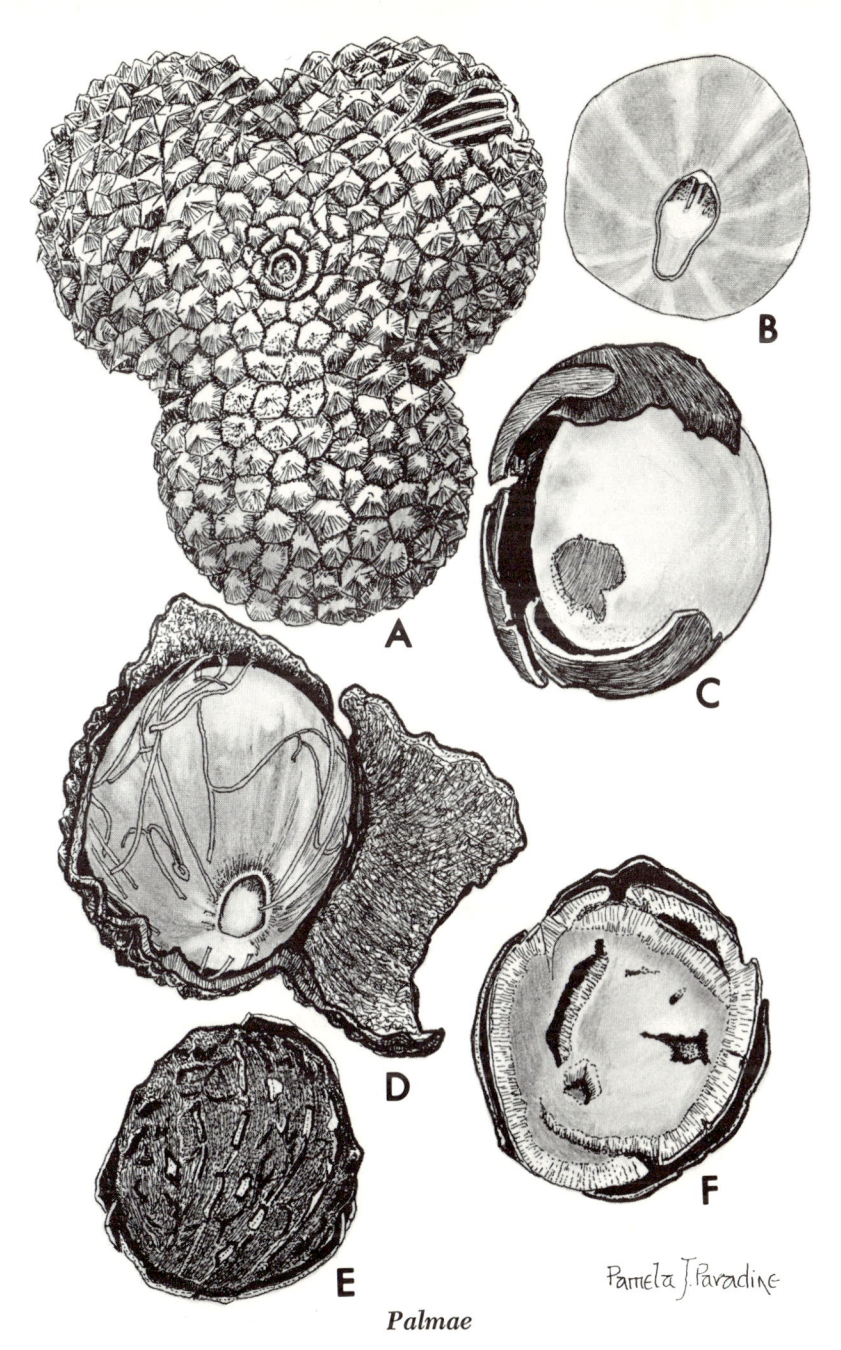

MANICARIA SACCIFERA GAERTNER

Figure 84. *Manicaria saccifera*. A, D, fruits; B-C, E-F, endocarps. A-B, D, basal views; C, E, lateral views; F, cross section (X1). Fruits and endocarps, Gunn and Dennis, southeastern coast of Florida.

Palmae

NYPA FRUTICANS WURMB., NYPA

Synonym: May be spelled *Nipa* in the drift literature.
Disseminule: Fruit.
Description: Fruit (A-B) 10 to 13 cm in diameter, round, compressed in cross section, black, surface deeply fluted from base to apex. The black paper-thin exocarp may be partially eroded revealing a fibrous mesocarp.
Buoyancy factor: Fibrous-corky mesocarp and small cavity within, or empty seed cavity.
Buoyancy: Not tested.
Viability: Not tested.
Currents: Tropical currents arising in southeastern Asia and Pacific regions.

Nypa, a native trunkless palm of the Indo-Malaysian region, has been spread by man to other tropical regions. The large pinnate leaves are often used as thatching for houses. Nypa trees often form extensive colonies in tidal basins and along tidal rivers. Both germinating and non-germinating fruits float, as well as rhizomes bearing young plants. There is no evidence to indicate that seeds remain viable for any appreciable length of time in seawater. Sea distribution may be of only local importance, even though the fruits are quite buoyant. Fossil fruits or casts have been found in the New World (Brazil, Mississippi, and Texas) and the Old World (Belgium, Egypt, France, Italy, and Russia) and New Zealand (Corner, 1966; Ridley, 1930). It is generally believed that these fruits represent drift material. While individual fruits float, there is no record that the aggregate fruits float as units. Each fruit is a member of a spherical aggregate fruit about 30 cm in diameter and composed of numerous individual fruits.

Palmae

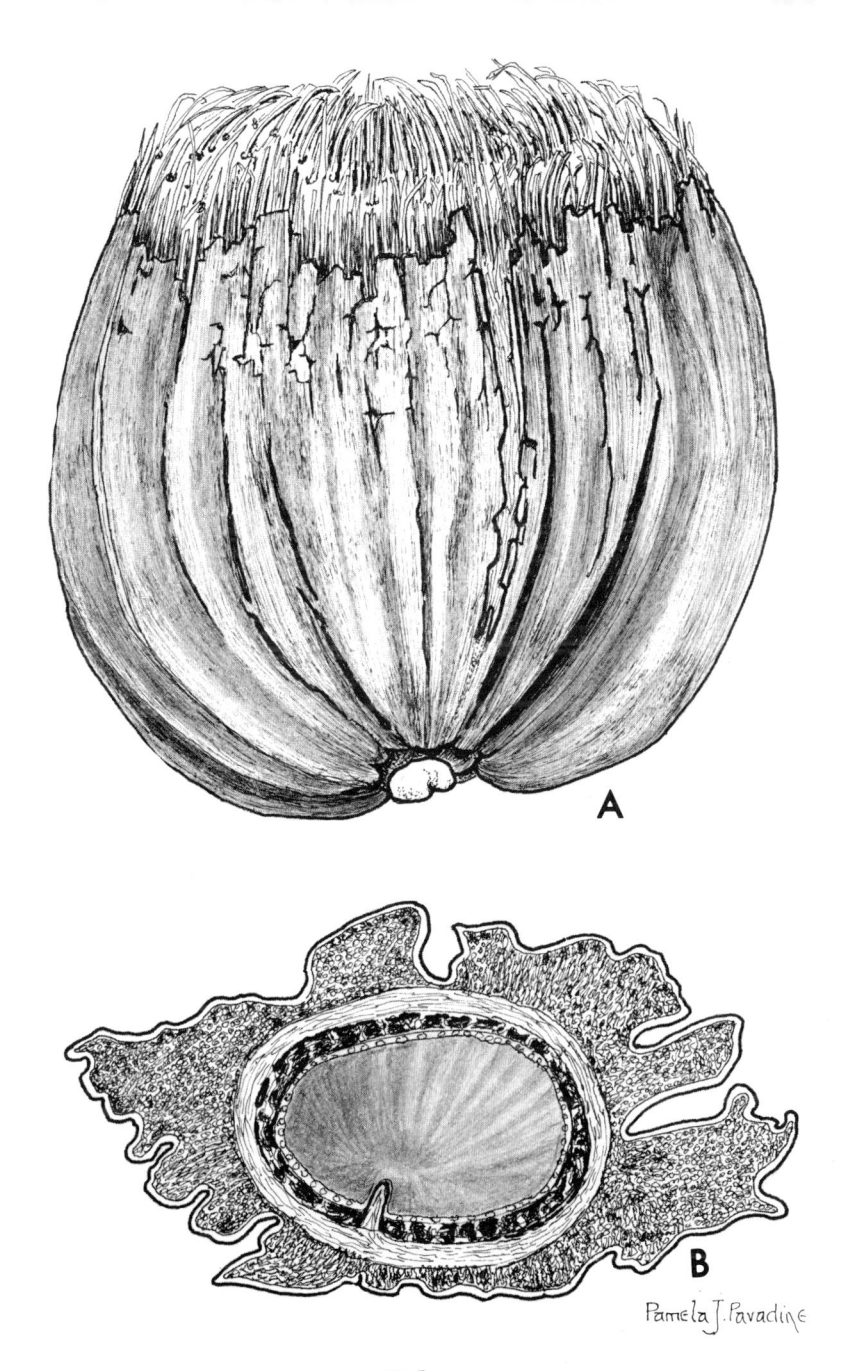

NYPA FRUTICANS WURMB., NYPA

Figure 85. *Nypa fruticans* fruits. A, lateral view; B, cross section (X1). Fruits, Degener and Degener, beaches of Canton Island.

Rhizophoraceae

RHIZOPHORA MANGLE L., RED MANGROVE

Disseminule: Seedling, rarely fruit.

Description: Seedling (Figs. 86 C-G and 87) up to 20 cm long, elongate, round in cross section, light to dark brown or grayish brown or mottled, basal tip often dark orange brown, surface smooth, rough, or ribbed. Some seedlings may bear roots (Fig. 87 A). Fruit (Fig. 86 A-B) 2.5 to 3.5 cm long, apex bearing a remnant stalk and calyx, base truncate and hollow.

Buoyancy factor: Corky seedling tissue.

Buoyancy: About 3 months.

Viability: Most seedlings viable.

Currents: Most tropical currents.

The red mangrove disseminule, often known as sea pencil, is unusual because it is a seedling, not a seed or fruit. The single seed germinates within the fruit while the fruit is still attached to the parent tree. After the seedling develops, an abscission layer forms and the seedling drops into the ocean or swamp where it may be automatically planted or washed away. This dispersal mechanism is quite effective, because the red mangrove is one of the most widely distributed species of the coastal tropical flora. Trees form colonies along the coast which may extend into the ocean. The colonies are best recognized by the stilt prop roots that rise above the soil and water level. Red mangrove colonies stabilize the coast and actively collect sand and silt to form new land. These colonies provide homes for various terrestrial and marine animals and epiphytic plants. Drifting seedlings transport marine fungi (Kohlmeyer, 1968). Johnston (1949) found seedlings of *R. samoensis* (Hochr.) Salvoza on the beaches of San Jose Island. These seedlings are similar to red mangrove seedlings. *Rhizophora samoensis* is a common coastal bushy mangrove whose natural range is the west coast of Central America, as well as the larger islands of eastern Polynesia. This distribution might be the result of drifting. Two other species, *Bruguiera gymnorrhiza* Lam. (Fig. 88 A) and *R. mucronulata* Lam. (Fig. 88 B-D), produce seedlings that drift in currents from east Africa to the Pacific region. Both seedlings are similar to the seedlings of red mangrove. Note the conspicuous calyx (strap-shaped lobes) on the *Bruguiera* seedling and the flask-shaped fruit on the *R. mucronulata* seedling.

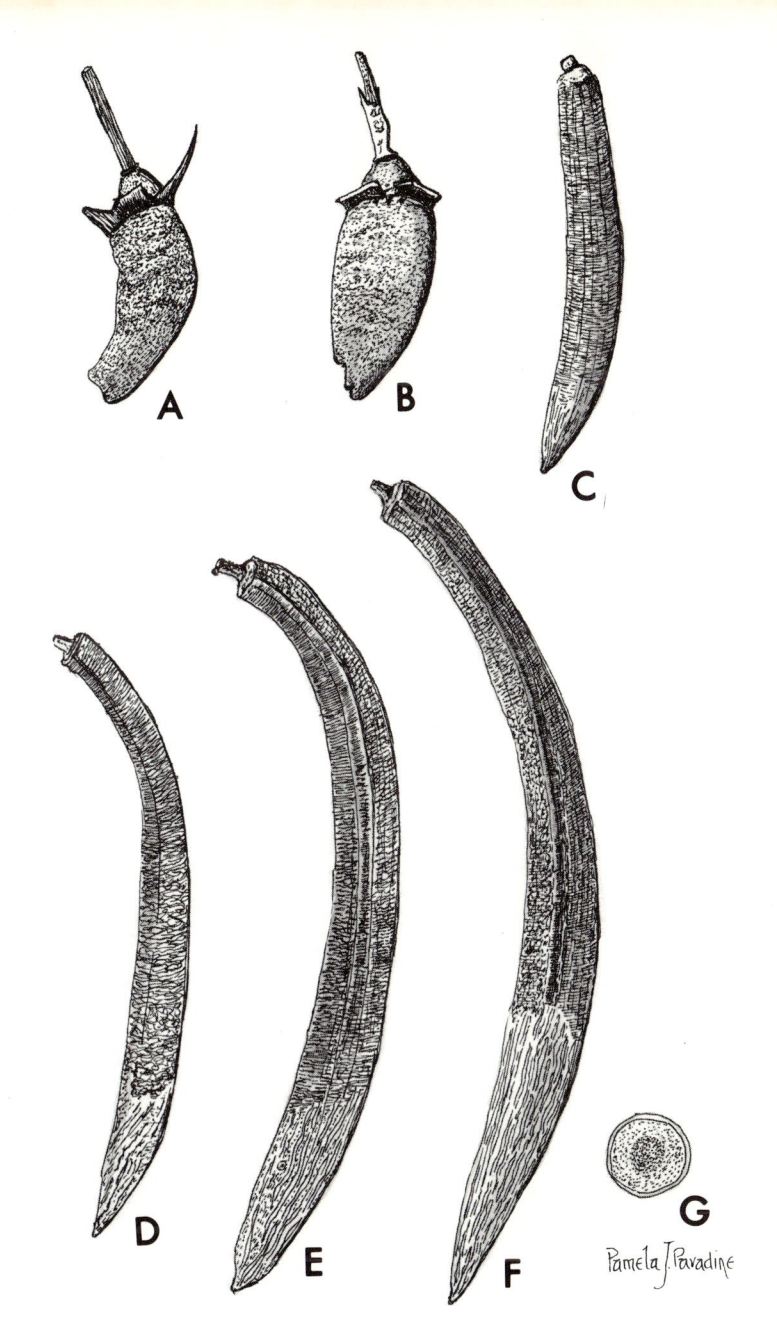

Rhizophoraceae

RHIZOPHORA MANGLE L. I

Figure 86. *Rhizophora mangle.* A-B, fruits; C-G, small seedlings. A-F, lateral views; G, cross section (X1). Fruits and seedlings, Gunn and Dennis, southeastern coast of Florida.

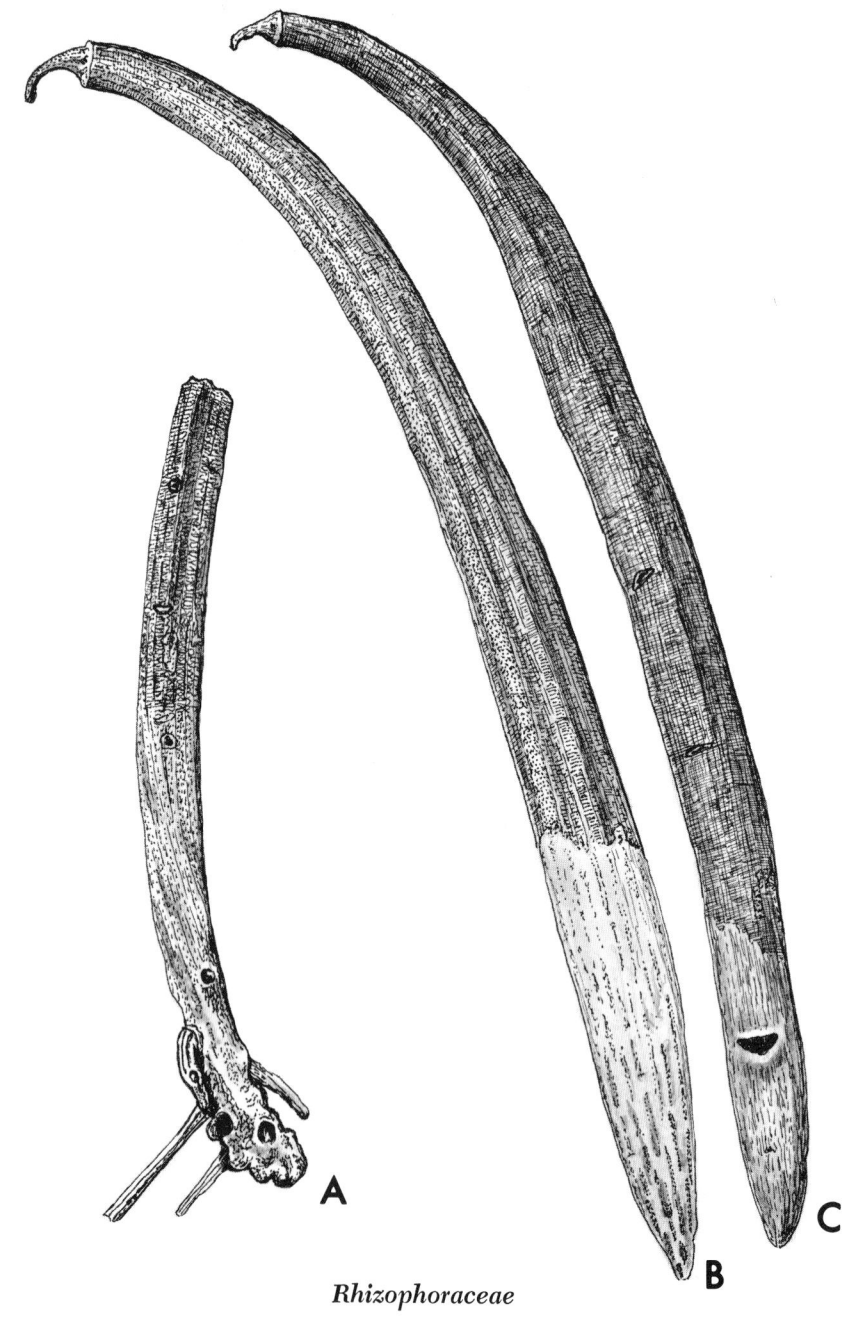

Rhizophoraceae

RHIZOPHORA MANGLE L. II

Figure 87. *Rhizophora mangle* seedlings. A, basal portion with roots; B, C, lateral views of mature seedlings (X1). Seedlings, Gunn and Dennis, southeastern coast of Florida.

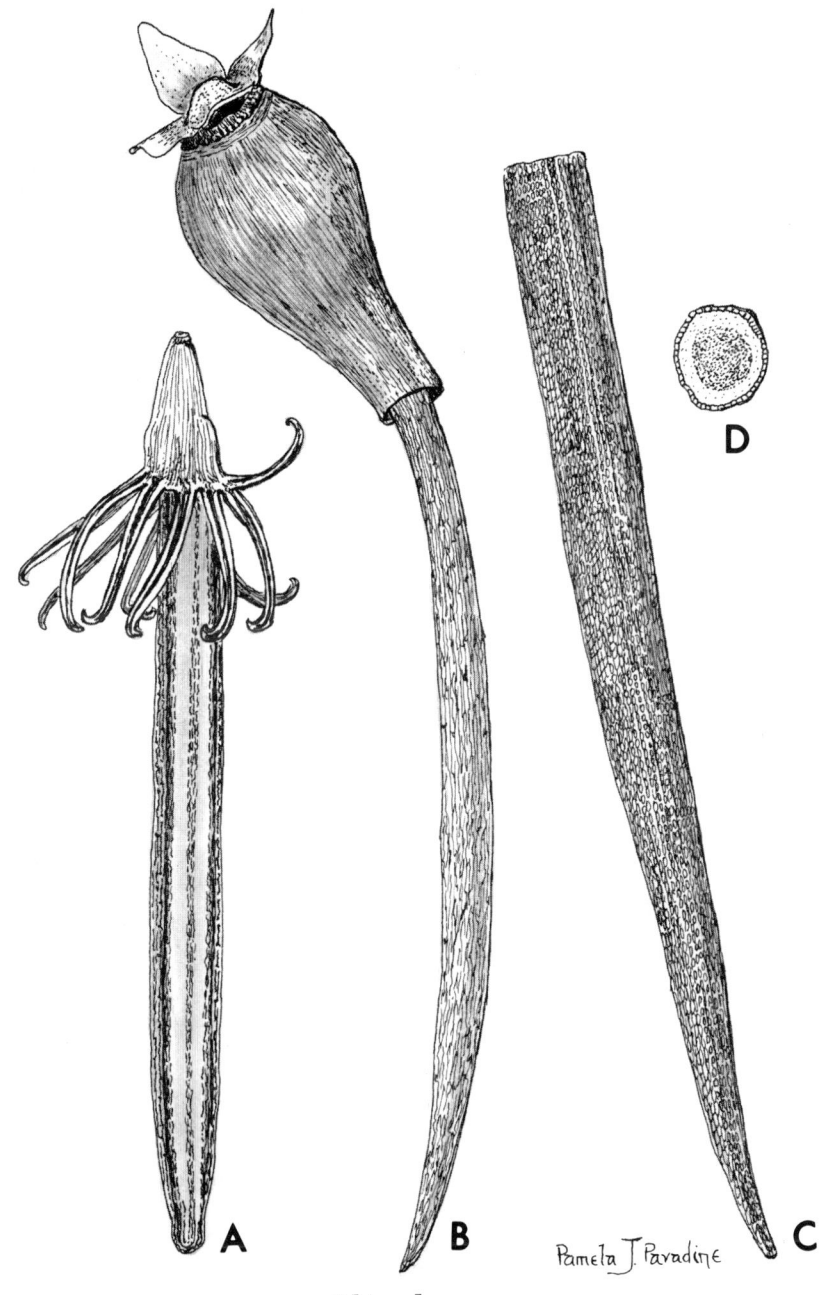

Rhizophoraceae

BRUGUIERA GYMNORRHIZA LAM.
RHIZOPHORA MUCRONULATA LAM.

Figure 88. *Bruguiera gymnorrhiza* (A); *Rhizophora mucronulata* (B-D). A-D, seedlings. A-B, lateral views; C, basal end in lateral view; D, cross section (X1). Fruits, from trees.

Sapindaceae

SAPINDUS SAPONARIA L., BLACK PEARL

Disseminule: Seed, rarely fruit.
Description: Seed (C-H) 10 to 15 mm in diameter, globose, round in cross section, lustrous black or brown, surface smooth. Fruit (A, B) up to 25 mm in diameter, surface black (yellow when fresh) and wrinkled, bearing a protruding basal scar.
Buoyancy factor: Space around embryo.
Buoyancy: About 14 months.
Viability: Most seeds viable.
Currents: Tropical currents arising in the New World.

Another common name is soapberry, because the mashed fruits, containing about 30 percent saponin, make suds in water. Drift seeds are known as black pearls, because they resemble black pearls and are used to make necklaces and as buttons or marbles. Neither the fruit nor the seed should be placed in the mouth. Crushed seeds are used as a fish poison, and the entire fruit is used as an astringent. While black pearl trees are native to the American tropics, they have been spread by man to the Old World tropics. Currents and birds are vectors in local colonization. Ridley (1930) reported that the first two plants known from the Bermuda Islands arose from drift piles. Guppy (1917) pointed out that drift seeds are not long lived in seawater, because of the weak hilar area. While he found seeds stranded on beaches of the Azores, seeds have not been recorded from European beaches. Seeds reaching Florida are usually viable, but they have their origin in Cuba. Heyerdahl and Ferdon (1961) cited the common name, parapara, in both Easter Island and the Americas as evidence that these two areas had prehistoric contact. This could also be used as evidence that the disseminules do not drift between these two regions.

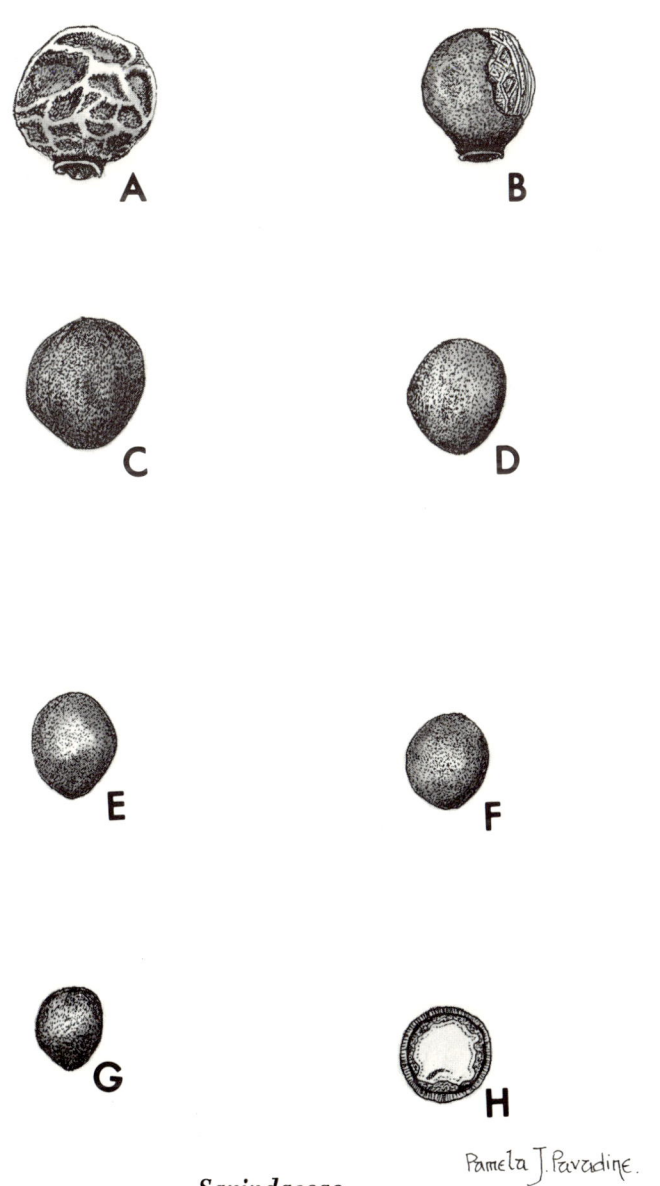

Sapindaceae

SAPINDUS SAPONARIA L.

Figure 89. *Sapindus saponaria.* A-B, fruits; C-H, seeds. A-G, lateral views; H, cross section (X1). Fruits and seeds, Gunn and Dennis, southeastern coast of Florida.

Sapotaceae

CALOCARPUM SPP., EGG FRUITS

Synonyms: *Lucuma* Molina and *Pouteria* Aublet in the drift literature.
Disseminule: Seed.
Description: Seed (B-G) 4 to 10 cm long, fusiform, elliptical, or obovoid, slightly compressed in cross section, shiny to dull dark brown to blackish, surface smooth and bearing a large (both in length and width) dull tan to light brown hilum which may have a slightly roughened surface. The hilum is shown in full face view in C. A portion of the black fruit coat remains attached to the left side of A.
Buoyancy factor: Empty space around embryo.
Buoyancy: Not tested.
Viability: One embryo tested, and it was viable.
Currents: Tropical currents arising in the New World.

The larger drift seeds (B-D) appear to be *C. mammosum* (L.) Cronquist, while the smaller ones (E-G) are members of one or more other species. Guppy (1917) reported finding two *Lucuma* (now *Calocarpum*) seeds on beaches of South Devon, England. We believe that the source of these seeds may have been a ship entering the English Channel. South Devon beaches are known for their beach garbage, because the shipping lanes are close to the beaches. The nearest recorded beach to England that has received a drift seed is the one at Roscarberry, County Cork, Ireland. The seed, deposited at National Museum of Ireland, Dublin under the genus name *Tieghemella* is probably *C. mammosum*. A similar seed was collected by Francis Z. Hough at Ocean City, Maryland in 1955. This is the only tropical disseminule that has been reported from a Maryland beach.

Most *Calocarpum* species are small to large trees which are native to tropical America. While the fruits of most species are edible, raw seeds of *C. mammosum* should not be eaten. Thoroughly cooked seeds are edible. Pittier (1914) reported an interesting use of *C. mammosum* seeds as irons to smooth starched white linen.

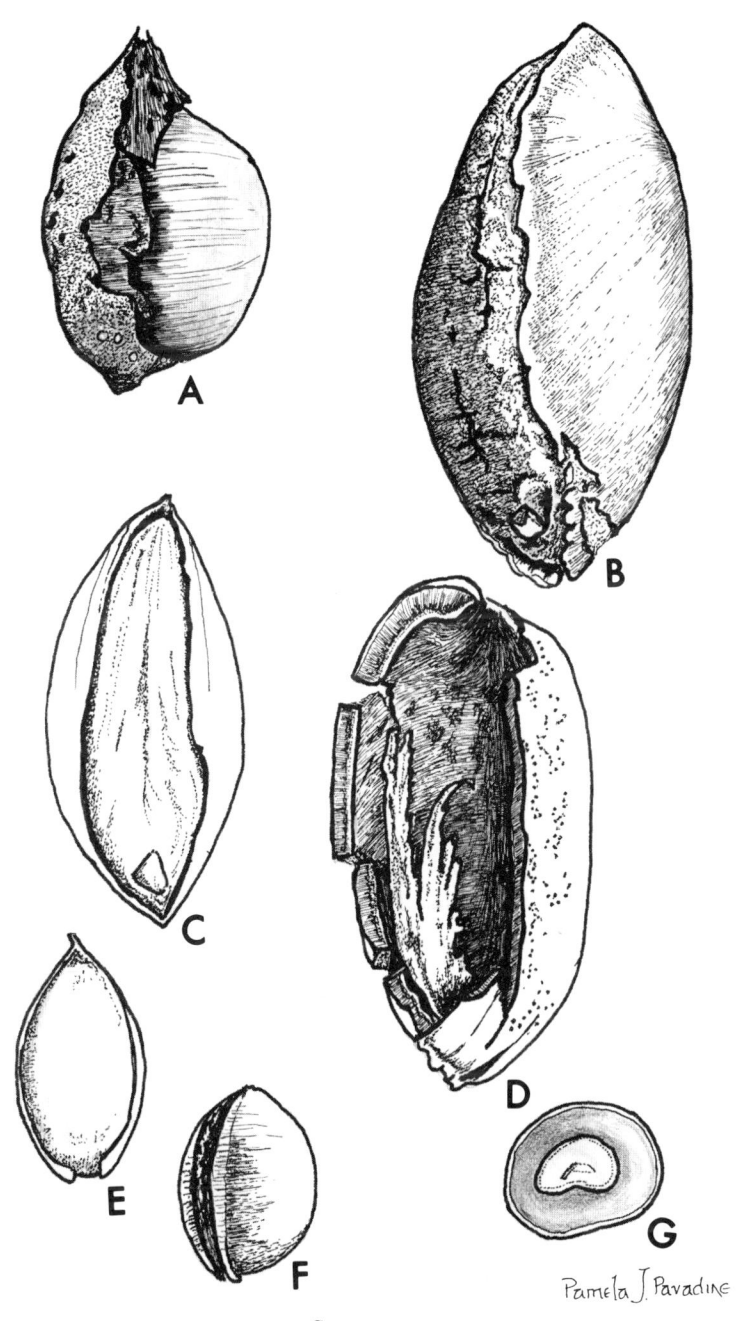

Sapotaceae

CALOCARPUM SPP.

Figure 90. *Calocarpum mammosum* (B-D); *C.* spp. (A, E-G). A, fruit partially eroded (present on left side); B-G, seeds. A-F, lateral views; G, cross section (X1). Fruit and seeds, Gunn and Dennis, southeastern coast of Florida.

Sapotaceae

MASTICHODENDRON CAPIRI (A. DC.) CRONQUIST VAR. TEMPISQUE (PITTIER) CRONQUIST

MASTICHODENDRON FOETIDISSIMUM (JACQ.) H. J. LAM

Synonym: *Sideroxylon* L. in the drift literature.
Disseminule: Seed.
Description: Seed of both species (A-I) 15 to 25 mm long, 10 to 17 mm in diameter, obovoid, round in cross section, lustrous brown to dull brown or blackish brown, surface smooth bearing a conspicuous light colored subterminal hilum.
Buoyancy factor: Space within seed.
Buoyancy: About 4 to 5 months.
Viability: None were found to be viable.
Currents: Tropical currents arising in the New World.

The seeds of at least two drift species are illustrated. The seeds are similar except for size. Seeds of *M. capiri* var. *tempisque* (A-C) are 20 to 25 mm long and 10 to 17 mm in diameter, while those of *M. foetidissimum* (D-L) are 15 to 17 mm long and 10 to 12 mm in diameter. Johnston (1949) observed that stranded seeds were empty. We have found that seeds stranded on Florida beaches were at least partially filled, though none were viable. *Mastichodendron capiri* var. *tempisque* is a native tree of Mexico and Central America, and *M. foetidissimum* is a native tree of Mexico and the Caribbean region where it inhabits coastal thickets and inland forests. Fruits of the latter species are sold in markets in Mexico. The yellow flesh has a bitter flavor.

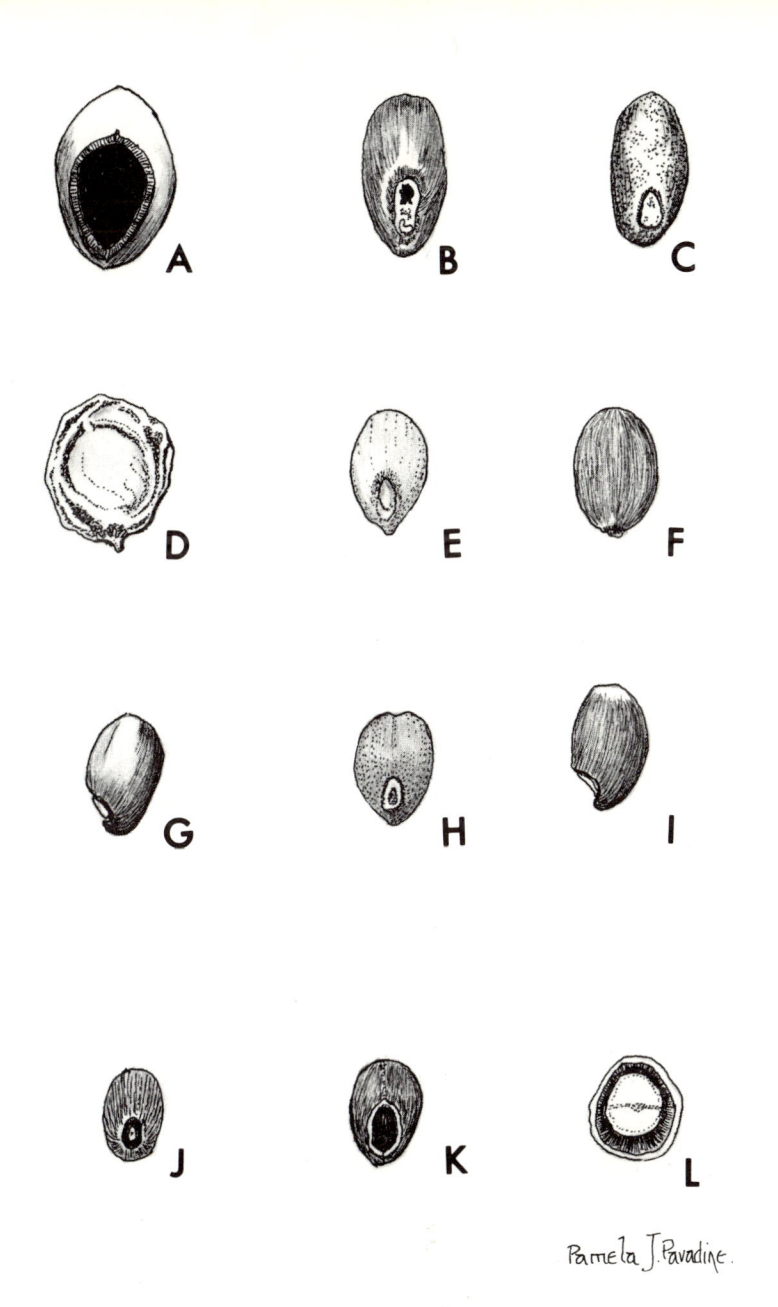

Sapotaceae

MASTICHODENDRON SPP.

Figure 91. *Mastichodendron capiri* var. *tempisque* (A-C); *M. foetidissimum* (D-L). A-C, E-L, seeds; A, K, damaged seeds; D, fruit. A-K, lateral views; L, cross section (X1). Seeds, Gunn and Dennis, southeastern coast of Florida; fruit, from tree.

(203)

Sterculiaceae

HERITIERA LITTORALIS DRYANDER, PUZZLE FRUIT

Disseminule: Fruit.
Description: Fruit (A-D) 4.5 to 8.5 cm long, ellipsoidal to oblong, slightly compressed in cross section, lustrous to dull light coppery to silver-gray brown to black, surface bearing a winglike keel which may be eroded (C).
Buoyancy factor: Fibrous corky mesocarp and space around seed.
Buoyancy: Not tested.
Viability: Not tested.
Currents: Tropical currents arising between East Africa and the Fiji Islands.

Another common name, looking-glass tree, is based on the silvery lower surfaces of the leaves which resemble mirrors. The tree is a native of southeastern Asia, and it inhabits tidal swamps where it often grows with *Calophyllum, Casuarina,* and *Rhizophora*. The present range of the puzzle fruit, the result of drifting, is from Madagascar to the Fiji Islands, and from India to Australia. The tree has not achieved a pantropic distribution by drifting. Ridley (1930) pointed out that it had not reached Hawaii, and Muir (1937) found only two fruits during his years of collecting stranded disseminules on the Riversdale beaches of South Africa.

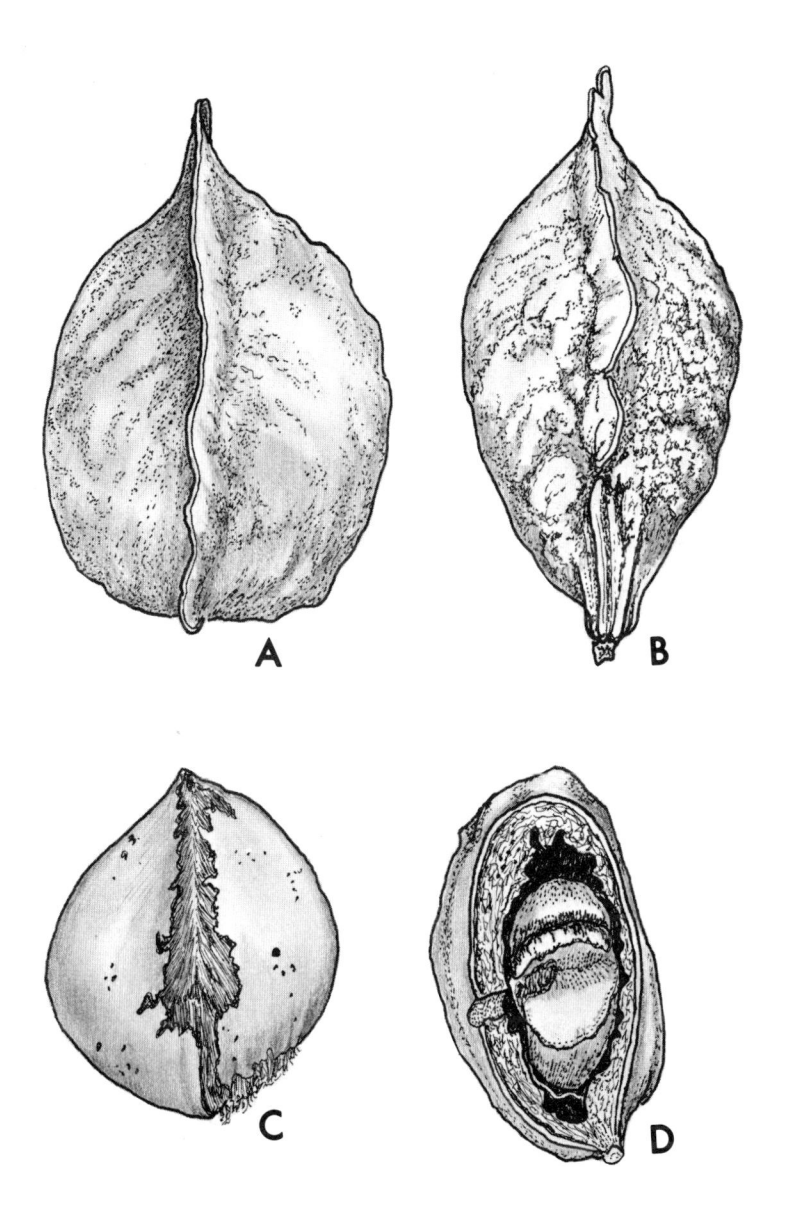

Sterculiaceae

HERITIERA LITTORALIS DRYANDER

Figure 92. *Heritiera littoralis* fruits. A-C, lateral views; D, longitudinal section (X1). Fruits, Degener and Degener, beaches of Canton Island.

Theaceae

PELLICIERA RHIZOPHORAE PLANCH & TRIANA

Disseminule: Fruit, rarely endocarp.
Description: Fruit (A-G) up to 11 cm long, up to 5 cm wide, usually longer than broad, ovate to round with apex prolonged into a stout beak up to 2.5 cm long, compressed in cross section, brown to blackish brown, surface corrugate. If the paper-thin exocarp is absent, the light tan endocarp is visible.
Buoyancy factor: Space around seed and an intercotyledonary cavity.
Buoyancy: Not tested.
Viability: Not tested.
Currents: Tropical currents along the Pacific coast of northern South America and Central America.

Pelliciera rhizophorae trees grow in mangrove swamps and attain a height of 6 to 15 m. The conic, fluted base of the trunk is about 3 times the diameter of the core of the trunk. This large base may prevent the trees from being blown over in the water-soaked soil. While we have found no evidence to indicate that these fruits are long distance drifters, they are effective local drifters. These fruits are especially adapted to drifting. The rounded basal end of the fruit is both the heaviest and the buoyant end. When the fruit falls from the tree, it hits the soil or water base first. When submerged in water, the fruit floats to the surface base up and beak down. The beak sheathes the well-developed embryonic root. Thus when drifting fruit is stranded, the beak tends to be "planted" in the mud or sand, facilitating the establishment of the seedling. Most fruits do not reach the beach, because the seeds germinate in seawater where the seedlings have little chance to survive.

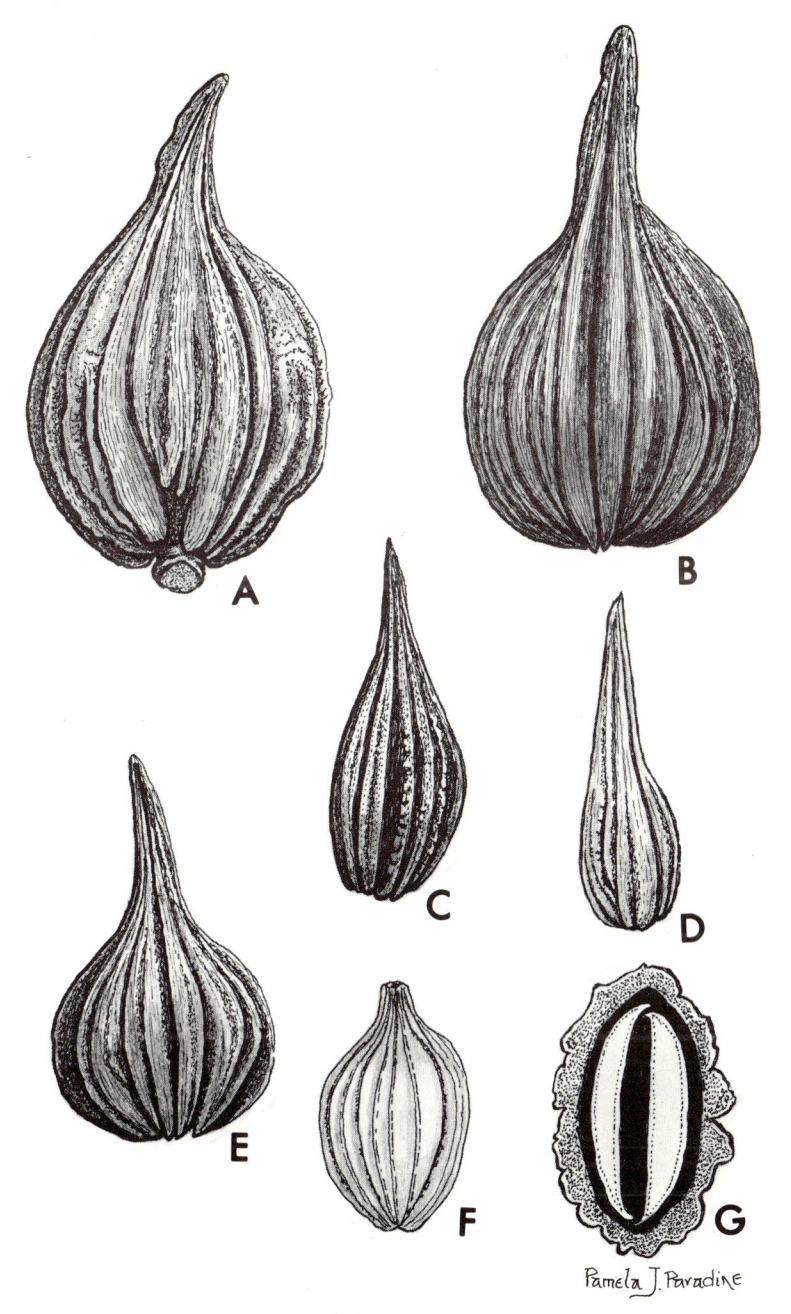

Theaceae

PELLICIERA RHIZOPHORAE PLANCH & TRIANA

Figure 93. *Pelliciera rhizophorae* fruits. A-F, lateral views; G, cross section (X1). Fruits, Edwards, beaches of Golfo de Nicoya, Costa Rica.

Appendix

Two lists of drift disseminules are presented, one alphabetically by genera, the other alphabetically by family name. These tropical disseminules float in seawater for at least one month. We consider them to be drift disseminules. An asterisk before the binomial indicates that the disseminule is illustrated in this book. The letters in parentheses are NW = New World, OW = Old World, and W = World. These letters indicate the primary area where the disseminules may be stranded. The notation beach garbage is applied to those disseminules whose presence on beaches is attributed to the activity of people rather than currents. The family name is noted for each binomial.

Generic List

Abuta sp. (NW)	Leguminosae
Acacia farnesiana Willd. (W)	Leguminosae
* *Acrocomia* spp. (NW)	Palmae
Aleurites fordii Hemsley (NW)	Euphorbiaceae
* *Aleurites moluccana* (L). Willd. (W)	Euphorbiaceae
Amygdalus persica L. (beach garbage)	Rosaceae
Anacardium occidentale L. (beach garbage)	Anacardiaceae
* *Andira galeottiana* Standley	Leguminosae
* *Andira inermis* (W. Wright) H. B. K. (NW)	Leguminosae
* *Annona glabra* L. (NW)	Annonaceae
* *Annona squamosa* L. (NW)	Annonaceae
* *Annona* sp. (NW)	Annonaceae
Arachis hypogaea L. (beach garbage)	Leguminosae
Arecastrum romanzoffiana Becc. (NW)	Palmae

Appendix

* *Astrocaryum* spp. (NW)	Palmae
* *Avicennia germinans* (L.) L. (W)	Avicenniaceae
Bactris balanoides (Oerst.) H. Wendl. (NW)	Palmae
* *Barringtonia asiatica* (L.) Kurz (OW)	Barringtoniaceae
Bertholletia excelsa Humb. & Bonpl. (NW)	Lecythidaceae
Blighia sapida Koenig (NW)	Sapindaceae
* *Borassus* spp. (OW)	Palmae
Brackenridgea sp. (OW)	Ochnaceae
* *Bruguiera gymnorrhiza* Lam. (OW)	Rhizophoraceae
* *Caesalpinia bonduc* (L.) Roxb. (W)	Leguminosae
* *Caesalpinia major* (Medikus) Dandy & Exell (NW)	Leguminosae
* *Caesalpinia* sp. (NW)	Leguminosae
Cakile spp. (W)	Cruciferae
* *Calatola costaricensis* Standley (NW)	Icacinaceae
* *Calocarpum mammosum* (L.) Cronquist (NW)	Sapotaceae
* *Calocarpum* spp. (NW)	Sapotaceae
Calodendrum capensis Thunb. (OW)	Rutaceae
* *Calophyllum calaba* L. (NW)	Guttiferae
* *Calophyllum inophyllum* L. (W)	Guttiferae
* *Canarium decumanum* Gaertner (OW)	Burseraceae
* *Canarium mehenbethune* Gaertner (OW)	Burseraceae
* *Canarium* sp. (OW)	Burseraceae
Canavalia bonariensis Lindley (NW)	Leguminosae
* *Canavalia cathartica* Thouars (OW)	Leguminosae
* *Canavalia nitida* (Cav.) Piper (NW)	Leguminosae
* *Canavalia rosea* (Sw.) DC. (W)	Leguminosae
Canavalia sericea A. Gray (OW)	Leguminosae
* *Canavalia* spp. (W)	Leguminosae
* *Carapa guianensis* Aublet (NW)	Meliaceae
* *Carya aquatica* (Michx. f.) Nutt. (NW)	Juglandaceae
* *Carya glabra* (Mill.) Sweet (NW)	Juglandaceae
* *Carya illinoensis* (Wang.) K. Koch (NW)	Juglandaceae
* *Carya tomentosa* Nutt. (NW)	Juglandaceae
* *Caryocar glabrum* (Aublet) Persoon (NW)	Caryocaraceae
* *Caryocar microcarpum* Ducke (NW)	Caryocaraceae
Caryocar villosum (Aublet) Persoon (NW)	Caryocaraceae
* *Cassia fistula* L. (NW)	Leguminosae
* *Cassia grandis* L. f. (NW)	Leguminosae
Cassytha filiformis L. (OW)	Lauraceae
Castanea sp. (NW)	Fagaceae

Appendix

* *Castanospermum australe* A. Cunn. (OW) — Leguminosae
* *Cerbera manghas* L. (OW) — Apocynaceae
* *Cerbera odollam* Gaertner (OW) — Apocynaceae
Ceriops tagal Robins (OW) — Rhizophoraceae
Chlaenandra sp. (OW) — Menispermaceae
Chrysobalanus icaco L. (NW) — Chrysobalanaceae
Chrysophyllum cainito L. (NW) — Sapotaceae
Citrullus lanatus (Thunb.) Mansf. (beach garbage) — Cucurbitaceae
Citrus spp. (W) — Rutaceae
Clerodendrum inerme Gaertner (OW) — Verbenaceae
* *Cocoid palms* (NW) — Palmae
* *Cocos nucifera* L. (W) — Palmae
Colubrina asiatica Brongn. (OW) — Rhamnaceae
Conocarpus erectus L. (NW) — Combretaceae
Cordia subcordata Lam. (OW) — Ehretiaceae
Corylus avellana L. (beach garbage) — Corylaceae
* *Crescentia cujete* L. (NW) — Bignoniaceae
* *Crinum americanum* L. (NW) — Amaryllidaceae
* *Crinum asiaticum* L. (OW) — Amaryllidaceae
Crudia schreberi J. F. Gmel. (NW) — Leguminosae
Cryptocarya latifolia Sond. (OW) — Lauraceae
* *Cycas circinalis* L. (W) — Cycadaceae
Cynometra cauliflora L. (OW) — Leguminosae

* *Dalbergia ecastaphyllum* (L.) Taub. (NW) — Leguminosae
* *Dalbergia monetaria* L. f. (NW) — Leguminosae
* *Delonix regia* (Hooker) Raf. (W) — Leguminosae
Dendrolobium umbellatum Bentham (OW) — Leguminosae
Dendrosicus latifolius (Miller) A. Gentry (NW) — Bignoniaceae
Derris trifoliata Lour. (OW) — Leguminosae
Dialium schlechteri Harms (OW) — Leguminosae
* *Dioclea megacarpa* Rolfe (NW) — Leguminosae
Dioclea panamensis Walp. (NW) — Leguminosae
* *Dioclea reflexa* Hooker f. (W) — Leguminosae
* *Dioclea* sp. (OW) — Leguminosae
Dodonaea viscosa L. (W) — Sapindaceae

Elaeis guineensis Jacq. (W) — Palmae
Elaeodendron xylocarpum DC. (NW) — Celastraceae
Enallagma latifolia (Mill.) Small (NW) — Bignoniaceae
* *Entada gigas* (L.) F. & R. (NW) — Leguminosae
* *Entada phaseoloides* (L.) Merrill (OW) — Leguminosae

Appendix

* *Enterolobium cyclocarpum* (Jacq.) Griseb. (NW)	Leguminosae
* *Enterolobium timbouva* Martius (NW)	Leguminosae
Erythrina fusca Lour. (OW)	Leguminosae
Erythrina variegata L. var. *orientalis* Merrill (W)	Leguminosae
* *Erythrina* spp. (W)	Leguminosae
Excoecaria indica Muell. Arg. (OW)	Euphorbiaceae
* *Fevillea cordifolia* L. (NW)	Cucurbitaceae
Ficus spp. (W)	Moraceae
Flindersia amboinensis Poir. (OW)	Meliaceae
Garcinia mangostana L. (beach garbage)	Guttiferae
Genipa clusiifolia Gr. (NW)	Rubiaceae
Gnetum sp. (OW)	Gnetaceae
Grevillea gibbosa R. Br. (OW)	Proteaceae
* *Grias cauliflora* L. (NW)	Lecythidaceae
Guazuma ulmifolia Lam. (NW)	Sterculiaceae
Guettardia speciosa L. (OW)	Rubiaceae
Gyrocarpus americanus Jacq. (W)	Gyrocarpaceae
Helinus ovata E. Mey. (OW)	Rhamnaceae
* *Heritiera littoralis* Dryander (OW)	Sterculiaceae
* *Hernandia nymphiifolia* (Presl) Kubitzi (OW)	Hernandiaceae
* *Hernandia sonora* L. (NW)	Hernandiaceae
Hevea brasiliensis Muell. Arg. (NW)	Euphorbiaceae
Hibiscus diversifolius Jacq. (W)	Malvaceae
Hibiscus tiliaceus L. (W)	Malvaceae
* *Hippomane mancinella* L. (W)	Euphorbiaceae
Hura crepitans L. (W)	Euphorbiaceae
* *Hymenaea courbaril* L. (NW)	Leguminosae
Inocarpus edulis Forst. (OW)	Leguminosae
* *Intsia bijuga* (Colebr.) O. Kuntze (OW)	Leguminosae
* *Ipomoea alba* L. (W)	Convolvulaceae
* *Ipomoea macrantha* R. & S. (W)	Convolvulaceae
* *Ipomoea pes-caprae* (L.) R. Br. (W)	Convolvulaceae
* *Ipomoea* spp. (W)	Convolvulaceae
Jatropha curcas L. (OW)	Euphorbiaceae
* *Juglans cinerea* L. (NW)	Juglandaceae
* *Juglans jamaicensis* C. DC. (NW)	Juglandaceae
* *Juglans nigra* L. (NW)	Juglandaceae
* *Juglans regia* L. (NW)	Juglandaceae

Appendix

Lactaria salubris Rumphius (OW)	Apocynaceae
Lecythis sp. (NW)	Lecythidaceae
* *Lodoicea maldivica* (Gmel.) Persoon (OW)	Palmae
Luffa insularum A. Gray (OW)	Cucurbitaceae
Lumnitzera racemosa Willd. (OW)	Combretaceae
Machaerium lunatum (G. F. W. Meyer) Ducke (NW)	Leguminosae
* *Mammea americana* L. (NW)	Guttiferae
* *Mangifera indica* L. (W)	Anacardiaceae
* *Manicaria saccifera* Gaertner (NW)	Palmae
* *Mastichodendron capiri* (A. DC.) Cronquist var. *tempisque* (Pitt.) Cronquist (NW)	Sapotaceae
* *Mastichodendron foetidissimum* (Jacq.) H. J. Lam (NW)	Sapotaceae
* *Maximiliana caribaea* Griseb. & Wendl. (NW)	Palmae
* *Merremia discoidesperma* (Donn. Sm.) O'Donell (NW)	Convolvulaceae
* *Merremia tuberosa* (L.) Rendle (W)	Convolvulaceae
* *Mora excelsa* Bentham (NW)	Leguminosae
* *Mora oleifera* (Triana) Ducke (NW)	Leguminosae
* *Mora* sp. (NW)	Leguminosae
Morinda spp. (W)	Rubiaceae
* *Mucuna fawcettii* Urban (NW)	Leguminosae
Mucuna flagellipes Vogel (OW)	Leguminosae
* *Mucuna gigantea* (Willd.) DC. (OW)	Leguminosae
Mucuna myriaptera Baker (OW)	Leguminosae
* *Mucuna nigricans* Steudel (OW)	Leguminosae
* *Mucuna sloanei* F. & R. (NW)	Leguminosae
* *Mucuna urens* (L.) Medikus (NW)	Leguminosae
* *Mucuna* spp. (W)	Leguminosae
Myristica fragrans Houtt. (W)	Myristicaceae
Myristica surinanensis Hooker f. & Thomson (NW)	Myristicaceae
Noltia africana Harv. & Sond. (OW)	Rhamnaceae
Noronhia emarginata Thouars (W)	Oleaceae
* *Nypa fruticans* Wurmb. (OW)	Palmae
Ochrosia elliptica Labill. (W)	Apocynaceae
* *Omphalea diandra* L. (NW)	Euphorbiaceae
* *Omphalea triandra* L. (NW)	Euphorbiaceae
Orania aruensis Becc. (OW)	Palmae

Appendix

* *Orbignya cohune* (Mart.) Standley (NW)	Palmae
* *Oxyrhynchus trinervius* (Donn. Sm.) Rudd (NW)	Leguminosae
Pachystigma spp. (W)	Rubiaceae
Palaquim sp. (OW)	Sapotaceae
Pandanus spp. (W)	Pandanaceae
* *Pangium edule* Reinw. (W)	Flacourtiaceae
Parinarium glaberrimum Hassk. (OW)	Rosaceae
Passiflora spp. (W)	Passifloraceae
* *Pelliciera rhizophorae* Planch & Triana (NW)	Theaceae
* *Peltophorum inerme* (Roxb.) Naves (NW)	Leguminosae
Persea americana Mill. (W)	Lauraceae
Physostigma cylindrosperma (Bak.) Homes (OW)	Leguminosae
Physostigma venenosum Balf. f. (OW)	Leguminosae
* *Phytelephas macrocarpa* Ruiz & Pav. (NW)	Palmae
Pinus spp. (NW)	Pinaceae
Platanus occidentalis L. (NW)	Platanaceae
Podocarpus sp. (NW)	Podocarpaceae
Pongamia pinnata Pierre (W)	Leguminosae
Prioria copaifera Griseb. (NW)	Leguminosae
Prunus spp. (NW)	Rosaceae
Psidium sp. (NW)	Myrtaceae
* *Pterocarpus officinalis* Jacq. (NW)	Leguminosae
* *Pterocarpus* sp. (NW)	Leguminosae
* *Quercus bennettii* Miq. (OW)	Fagaceae
* *Quercus* spp. (NW)	Fagaceae
* *Rhizophora mangle* L. (W)	Rhizophoraceae
* *Rhizophora mucronulata* Lam. (OW)	Rhizophoraceae
Rhizophora samoensis (Hochr.) Salvoza (W)	Rhizophoraceae
Ricinus communis L. (W)	Euphorbiaceae
* *Sacoglottis amazonica* Martius (NW)	Humiriaceae
Sacoglottis gabonensis Urban (OW)	Humiriaceae
* *Sapindus saponaria* L. (NW)	Sapindaceae
Scaevola koenigii Vahl (OW)	Goodeniaceae
Scaevola plumieri Vahl (W)	Goodeniaceae
Scyphiphora hydrophylacea Gaertner f. (OW)	Rubiaceae
Smilax bona-nox L. tubers (NW)	Liliaceae
Sonneratia caseolaris Druce (OW)	Sonneratiaceae
Sophora tomentosa L. (W)	Leguminosae

Appendix

* *Spondias dulcis* S. Parkinson (OW) — Anacardiaceae
* *Spondias mombin* L. (NW) — Anacardiaceae
Sterculia carthaginensis Cav. (NW) — Sterculiaceae
* *Strongylodon lucidus* (Forst. f.) Seem. (OW) — Leguminosae
Suriana maritima L. (OW) — Simaraubaceae
* *Swietenia mahagoni* (L.) Jacq. (NW) — Meliaceae

Tacca leoteopetaloides (L.) O. Kuntze (OW) — Taccaceae
Tacca pinnatifida Forst. (OW) — Taccaceae
Telfairia pedata Hooker (OW) — Cucurbitaceae
* *Terminalia catappa* L. (W) — Combretaceae
* *Terminalia* spp. (OW) — Combretaceae
Theobroma cacao L. (W) — Sterculiaceae
Thespesia populnea (L.) Correa (W) — Malvaceae
Tournefortia argentea L. f. (OW) — Boraginaceae
Tournefortia gnaphalodes R. Br. (NW) — Boraginaceae
Trapa bispinosa Roxb. (OW) — Hydrocaryaceae
Triumfetta procumbens Forst. (OW) — Tiliaceae

* *Vantanea guianensis* Aublet (NW) — Humiriaceae
Vateria papuana Dyer (OW) — Dipterocarpaceae
Vigna lutea A. Gray (OW) — Leguminosae
Vigna luteola Bentham (NW) — Leguminosae

Wedelia biflora DC. (OW) — Compositae

Ximenia americana L. (W) — Oleaceae
Xylocorpus moluccensis (Lam.) Roem. (OW) — Meliaceae

❧ Family List

Amaryllidaceae
    * *Crinum americanum* L. (NW)
    * *Crinum asiaticum* L. (OW)
Anacardiaceae
    Anacardium occidentale L. (beach garbage)
    * *Mangifera indica* L. (W)
    * *Spondias dulcis* Parkinson (OW)
    * *Spondias mombin* L. (NW)
Annonaceae
    * *Annona glabra* L. (NW)

Appendix

    * *Annona squamosa* L. (NW)
    * *Annona* sp. (NW)
Apocynaceae
    * *Cerbera manghas* L. (OW)
    * *Cerbera odollam* Gaertner (OW)
    Lactaria salubris Rumphius (OW)
    Ochrosia elliptica Labill. (W)
Avicenniaceae
    * *Avicennia germinans* (L.) L. (W)

Barringtoniaceae
    * *Barringtonia asiatica* (L.) Kurz (OW)
Bignoniaceae
    * *Crescentia cujete* L. (NW)
    * *Dendrosicus latifolius* (Miller) A. Gentry (NW)
    Enallagma latifolia (Mill.) Small (NW)
Boraginaceae
    Tournefortia argentea L. f. (OW)
    Tournefortia gnaphalodes R. Br. (NW)
Burseraceae
    * *Canarium decumanum* Gaertner (OW)
    * *Canarium mehenbethune* Gaertner (OW)
    * *Canarium* sp. (OW)

Caryocaraceae
    * *Caryocar glabrum* (Aublet) Persoon (NW)
    * *Caryocar microcarpum* Ducke (NW)
    Caryocar villosum (Aublet) Persoon (NW)
Casuarinaceae
    Casuarina equisetifolia L. (W)
Celastraceae
    Elaeodendron xylocarpum DC. (NW)
Chrysobalanaceae
    Chrysobalanus icaco L. (NW)
Combretaceae
    Conocarpus erectus L. (NW)
    Lumnitzera racemosa Willd. (OW)
    * *Terminalia catappa* L. (W)
    * *Terminalia* spp. (OW)
Compositae
    Wedelia biflora DC. (OW)
Convolvulaceae
    * *Ipomoea alba* L. (W)

Appendix

    * *Ipomoea macrantha* R. & S. (W)
    * *Ipomoea pes-caprae* (L.) R. Br. (W)
    * *Ipomoea* spp. (W)
    * *Merremia discoidesperma* (Donn. Sm.) O'Donell (NW)
    * *Merremia tuberosa* (L.) Rendle (W)
Corylaceae
    Corylus avellana L. (beach garbage)
Cruciferae
    Cakile spp. (W)
Cucurbitaceae
    Citrullus lanatus (Thunb.) Mansf. (beach garbage)
    * *Fevillea cordifolia* L. (NW)
    Luffa insularum A. Gray (OW)
    Telfairia pedata Hooker (OW)
Cycadaceae
    * *Cycas circinalis* L. (W)

Dipterocarpaceae
    Vateria papuana Dyer (OW)

Ehretiaceae
    Cordia subcordata Lam. (OW)
Euphorbiaceae
    Aleurites fordii Hemsley (NW)
    * *Aleurites moluccana* (L.) Willd. (W)
    Excoecaria indica Muell. Arg. (OW)
    Hevea brasiliensis Muell. Arg. (NW)
    * *Hippomane mancinella* L. (W)
    Hura crepitans L. (W)
    Jatropha curcas L. (OW)
    * *Omphalea diandra* L. (NW)
    * *Omphalea triandra* L. (NW)
    Ricinus communis L. (W)

Fagaceae
    Castanea sp. (NW)
    * *Quercus bennettii* Miq. (OW)
    * *Quercus* spp. (NW)
Flacourtiaceae
    * *Pangium edule* Reinw. (W)

Gnetaceae
    Gnetum sp. (OW)

Appendix

Goodeniaceae
    Scaevola koenigii Vahl (OW)
    Scaevola plumieri Vahl (W)
Guttiferae
    * *Calophyllum calaba* L. (NW)
    * *Calophyllum inophyllum* L. (W)
    Garcinia mangostana L. (beach garbage)
    * *Mammea americana* L. (NW)
Gyrocarpaceae
    Gyrocarpus americanus Jacq. (W)

Hernandiaceae
    * *Hernandia nymphiifolia* (Presl) Kubitzi (OW)
    * *Hernandia sonora* L. (NW)
Humiriaceae
    * *Sacoglottis amazonica* Martius (NW)
    Sacoglottis gabonensis Urban (OW)
    * *Vantanea guianensis* Aublet (NW)
Hydrocaryaceae
    Trapa bispinosa Roxb. (OW)

Icacinaceae
    * *Calatola costaricensis* Standley (NW)

Juglandaceae
    * *Carya aquatica* (Michx. f.) Nutt. (NW)
    * *Carya glabra* (Mill.) Sweet (NW)
    * *Carya illinoensis* (Wang.) K. Koch (NW)
    * *Carya tomentosa* Nutt. (NW)
    * *Juglans cinerea* L. (NW)
    * *Juglans jamaicensis* C. DC. (NW)
    * *Juglans nigra* L. (NW)
    * *Juglans regia* L. (NW)

Lauraceae
    Cassytha filiformis L. (OW)
    Cryptocarya latifolia Sond. (OW)
    Persea americana Mill. (W)
Lecythidaceae
    Bertholletia excelsa Humb. & Bonpl. (NW)
    * *Grias cauliflora* L. (NW)
    Lecythis sp. (NW)

Appendix

Leguminosae
    Abuta sp. (NW)
    Acacia farnesiana Willd. (W)
    * *Andira galeottiana* Standley
    * *Andira inermis* (W. Wright) H. B. K. (NW)
    Arachis hypogaea L. (beach garbage)
    * *Caesalpinia bonduc* (L.) Roxb. (W)
    * *Caesalpinia major* (Medikus) Dandy & Exell (NW)
    * *Caesalpinia* sp. (NW)
    Canavalia bonariensis Lindley (NW)
    * *Canavalia cathartica* Thouars (OW)
    * *Canavalia nitida* (Cav.) Piper (NW)
    * *Canavalia rosea* (Sw.) DC. (W)
    Canavalia sericea A. Gray (OW)
    * *Canavalia* spp. (W)
    * *Cassia fistula* L. (NW)
    * *Cassia grandis* L. f. (NW)
    * *Castanospermum australe* A. Cunn. (OW)
    Crudia schreberi J. F. Gmel. (NW)
    Cynometra cauliflora L. (OW)
    * *Dalbergia ecastaphyllum* (L.) Taub. (NW)
    * *Dalbergia monetaria* L. f. (NW)
    * *Delonix regia* (Hooker) Raf. (W)
    Dendrolobium umbellatum Bentham (OW)
    Derris trifoliata Lour. (OW)
    Dialium schlechteri Harms (OW)
    * *Dioclea megacarpa* Rolfe (NW)
    Dioclea panamensis Walp. (NW)
    * *Dioclea reflexa* Hooker f. (W)
    * *Dioclea* sp. (OW)
    * *Entada gigas* (L.) F. & R. (NW)
    * *Entada phaseoloides* (L.) Merrill (OW)
    * *Enterolobium cyclocarpum* (Jacq.) Griseb. (NW)
    * *Enterolobium timbouva* Martius (NW)
    Erythrina fusca Lour. (OW)
    Erythrina variegata L. var. *orientalis* (L.) Merrill
    * *Erythrina* spp. (W)
    * *Hymenaea courbaril* L. (NW)
    Inocarpus edulis Forst. (OW)
    * *Intsia bijuga* (Colebr.) O. Kuntze (OW)
    Machaerium lunatus (G. F. W. Meyer) Ducke (NW)
    * *Mora excelsa* Bentham (NW)

Appendix

    * *Mora oleifera* (Triana) Ducke (NW)
    * *Mora* sp. (NW)
    * *Mucuna fawcettii* Urban (NW)
    Mucuna flagellipes Vogel (OW)
    * *Mucuna gigantea* DC. (OW)
    Mucuna myriaptera Baker (OW)
    * *Mucuna nigricans* Steudel (OW)
    * *Mucuna sloanei* F. & R. (NW)
    * *Mucuna urens* (L.) Medikus (NW)
    * *Mucuna* spp. (W)
    * *Oxyrhynchus trinervius* (Donn. Sm.) Rudd (NW)
    * *Peltophorum inerme* (Roxb.) Naves (NW)
    Physostigma cylindrosperma (Bak.) Homes (OW)
    Physostigma venenosum Balf. f. (OW)
    Pongamia pinnata Pierre (W)
    Prioria copaifera Griseb. (NW)
    * *Pterocarpus officinalis* Jacq. (NW)
    * *Pterocarpus* sp. (NW)
    Sophora tomentosa L. (W)
    * *Strongylodon lucidus* (Forst. f.) Seem. (OW)
    Vigna lutea A. Gray (OW)
    Vigna luteola Bentham (NW)

Liliaceae
    Smilax bona-nox L. tubers (NW)

Malvaceae
    Hibiscus diversifolius Jacq. (W)
    Hibiscus tiliaceus L. (W)
    Thespesia populnea (L.) Correa (W)

Meliaceae
    * *Carapa guianensis* Aublet (NW)
    * *Carapa moluccensis* Lam. (OW)
    Cedrela odorata L. (NW)
    Flindersia amboinensis Poir. (OW)
    * *Swietenia mahagoni* (L.) Jacq. (NW)
    * *Xylocorpus moluccensis* (Lem.) Roem. (OW)

Menispermaceae
    Chlaenandra sp. (OW)

Moraceae
    Ficus spp. (NW)

Myristicaceae
    Myristica fragrans Houtt. (W)
    Myristica surinanensis Hooker f. & Thomson (NW)

Appendix

Myrtaceae
    Psidium sp. (NW)

Ochnaceae
    Brackenridgea sp. (OW)
Oleaceae
    Noronhia emarginata Thouars (W)
    Ximenia americana L. (W)

Palmae
    * *Acrocomia* spp. (NW)
    Arecastrum romanzoffiana Becc. (NW)
    * *Astrocaryum* spp. (NW)
    Bactris balanoides (Oerst.) H. Wendl. (NW)
    * *Borassus* spp. (OW)
    * *Cocoid palms* (NW)
    * *Cocos nucifera* L. (W)
    Elaeis guineensis Jacq. (W)
    * *Lodoicea maldivica* (Gmel.) Persoon (OW)
    * *Manicaria saccifera* Gaertner (NW)
    * *Maximiliana caribaea* Griseb. & Wendl.
    * *Nypa fruticans* Wurmb. (OW)
    Orania aurensis Becc. (OW)
    * *Orbignya cohune* (Mart.) Standley (NW)
    * *Phytelephas macrocarpa* Ruiz & Pav. (NW)
Pandanaceae
    Pandanus spp. (W)
Passifloraceae
    Passiflora spp. (W)
Pinaceae
    Pinus spp. (NW)
Platanaceae
    Platanus occidentalis L. (NW)
Podocarpaceae
    Podocarpus sp. (NW)
Proteaceae
    Grevillea gibbosa R. Br. (OW)

Rhamnaceae
    Colubrina asiatica Brongn. (OW)
    Helinus ovata E. Mey. (OW)
    Noltia africana Harv. & Sond. (OW)

Appendix

Rhizophoraceae
    * *Bruguiera gymnorrhiza* Lam. (OW)
    Ceriops tagal Robins (OW)
    * *Rhizophora mangle* L. (W)
    * *Rhizophora mucronulata* Lam. (OW)
    Rhizophora samoensis (Hochr.) Salvoza (W)
Rosaceae
    Amygdalus persica L. (beach garbage)
    Parinarium glaberrimum Hassk. (OW)
    Prunus spp. (NW)
Rubiaceae
    Genipa clusiifolia Gr. (NW)
    Guettardia speciosa L. (OW)
    Morinda spp. (W)
    Pachystigma spp. (W)
    Scyphiphora hydrophylacea Gaertner f. (OW)
Rutaceae
    Calodendrum capensis Thunb. (OW)
    Citrus spp. (W)

Sapindaceae
    Blighia sapida Koenig (NW)
    Dodonaea viscosa L. (W)
    Melicoccus bijugatus Jacq. (NW)
    * *Sapindus saponaria* L. (NW)
Sapotaceae
    * *Calocarpum mammosum* (L.) Cronquist (NW)
    Calocarpum spp. (NW)
    Chrysophyllum cainito L. (NW)
    * *Mastichodendron capiri* (A. DC.) Cronquist var. *tempisque* (Pitt.) Cronquist (NW)
    * *Mastichodendron foetidissimum* (Jacq.) H. J. Lam (NW)
    Palaquim sp. (OW)
Simaraubaceae
    Suriana maritima L. (OW)
Sonneratiaceae
    Sonneratia caseolaris Druce (OW)
Sterculiaceae
    Guazuma ulmifolia Lam. (NW)
    * *Heritiera littoralis* Dryander (OW)
    Sterculia carthaginensis Cav. (NW)
    Theobroma cacao L. (W)

Appendix

Taccaceae
    Tacca leoteopetaloides (L.) O. Kuntze (OW)
    Tacca pinnatifida Forst. (OW)
Theaceae
    * *Pelliciera rhizophorae* Planch & Triana (NW)
Tiliaceae
    Triumfetta procumbens Forst. (OW)

Verbenaceae
    Clerodendrum inerme Gaertner (OW)

Glossary

Abscission. Natural separation of a fruit from a plant by a special layer.
Carpel. A section of a fruit, such as a section of an orange.
Complete fruit. An entire fruit.
Cotyledon. Seed or primary leaf or leaves of an embryo.
Dehiscent. Fruit opening by 1 or more sutures, like a pea pod.
Disseminule. A collective term for true seeds, one-seeded fruits, fruits, and seedlings.
Drifter. A disseminule that floats in seawater for at least 1 month.
Embryo. Rudimentary plant within a seed.
Embryonic axis. The embryonic root and first true leaves or bud without the cotyledons.
Endocarp. Inner bony fruit layer (not always present).
Fruit. Seed-bearing organ; a ripened ovary.
Haustorium. An organ to absorb the endosperm.
Hilum. Scar on the seed coat showing where seed was attached to the fruit.
Incomplete fruit. A partial fruit, such as an endocarp or mesocarp.
Indehiscent. Fruit not opening by sutures, like an apple which must be cut open or rot to release its seeds.
Interstice. Space between ribs.
Littoral. A coast region, or referring to a shore of an ocean.
Mesocarp. Corky or fibrous layer of a fruit (not always present).
Panicle. An indeterminate much-branched arrangement of fruits.
Pantropic. Throughout the tropics.
Pores. One to 3 small round shallow pits or scars at the base of cocoid palm endocarps.
Sea-bean. Any tropical drift disseminule, often a legume seed.

Glossary

Seed. The essential parts are an embryo and 1 or 2 seed coats, and often a food reserve; a ripened ovule within an ovary or fruit.
Seedling. The germinating seed.
Spermatophyte. Seed-bearing plant.
Stone cells. Cells which have hardened surfaces like those of an endocarp.
Strand. A beach.
Stranded. Beached, as in a beached buoyant disseminule.
Suture. A line or mark of splitting open.
True sea-bean. A seed of 1 of 3 species of *Mucuna* (*M. fawcettii*, *M. sloanei*, or *M. urens*).
Truncate. Appearing to have been cut off.
Tuberculate. Bearing small rounded protruding bodies.

Bibliography

Adams, C. D., G. R. Proctor, and R. W. Read. 1972. Flowering plants of Jamaica. 848 pp. MacLehose, Glasgow.
Arnold, H. L. 1968. Poisonous plants of Hawaii. 71 pp. Tuttle, Rutland, Vermont.
Bates, H. W. 1863. Naturalist on the river Amazon. 2: (cf. pp. 170–171, 417). Murray, London.
Blake, H. 1825. Letters from the Irish Highlands. 367 pp. (cf. pp. 315–321). Murray, London.
Blatter, E. 1926. The palms of British India and Ceylon. 600 pp. (cf. pp. 213–245). Oxford University Press.
Brown, R. 1818. Appendix V. In: J. K. Tuckey. Narrative of an expedition to explore the river Zaire. 498 pp. (cf. pp. 481–482). Murray, London.
Candolle, A. L. P. P. de. 1855. Géographie botanique raisonnée. 2: 607–1365. Masson, Paris.
Carlquist, S. 1965. A natural history of the islands of the world. 451 pp. Natural History Press, Garden City, N.Y.
Carlquist, S. 1970. Hawaii, a natural history. 463 pp. (cf. pp. 102–105). Natural History Press, Garden City, N.Y.
Clough, P. 1969. The voyage and untimely death of *Mucuna urens*. Gard. Chron. & New Hort. 165 (23): 4.
Clusius, C. 1605. Exoticorum libri decem. 724 pp. Rapheleng, Leiden.
Coe, C. H. 1894. So-called Florida sea-beans. Garden and Forest, no. 356: 502–504.
Colgan, N. 1919. On the occurrence of tropical drift seeds on the Irish Atlantic coasts. Roy. Irish Proc. 35 (B3): 29–54; plates.
Cooper, R. C. 1967. *Ipomoea pes-caprae* (Convolvulaceae) on Ninety Mile Beach, New Zealand. Rec. Auck. Inst. Mus. 6: 171–174.

Bibliography

Corner, E. J. H. 1966. Natural history of palms. 386 pp. University of California Press, Berkeley.

Cuatrecasas, J. 1961. A taxonomic revision of the Humiriaceae. Contri. U.S. Nat. Herb. *35*: 25–214.

Dahlgren, B. E. and P. C. Standley. 1944. Edible and poisonous plants of the Caribbean region. 102 pp. U.S. Navy Department, Bureau of Medicine and Surgery, Washington.

Darwin, C. 1855. Effect of salt-water on germination of seeds. Gard. Chron. & Agri. Gaz. *1855* (47): 773.

Darwin, C. 1857. On action of sea-water on germination of seeds. Jour. Linn. Soc. *1*: 130–140.

Darwin, C. 1883. Journal of researches. 519 pp. (Chapter 20, a reprint of an article published in 1836). Appleton, New York.

Degener, O. 1945. Plants of Hawaii national parks illustrative of plants and customs of the South Seas. 310 pp. Edwards, Ann Arbor.

Dennis, J. V. and C. R. Gunn. 1971. Case against trans-Pacific dispersal of the coconut by ocean currents. Econ. Bot. *25*: 407–413.

Dennis, J. V. and C. R. Gunn. 1975. Sea-beans from beaches of Cape Cod and the islands. Cape Naturalist. *3* (3): 40–45.

English, T. M. S. 1913. Some notes from a West Indian coral island. Bull. Misc. Info. *10*: 367–372.

Erslev, E. 1877. Mimosefrø, fundne i Jylland. Geografisk Tidskrift *1*: 79–80.

Fairchild, D. 1931. Exploring for plants. 591 pp. (cf. pp. 332; plate). Macmillan, New York.

Fairchild, D. 1943. Garden islands of the great east. 239 pp. Scribner, New York.

Gaertner, J. 1789–1790. Fructibus et seminibus plantarum. Plate volumes. Author, Stuttgart.

Gibbons, E. 1967. Beachcomber's handbook. 230 pp. (cf. p. 132). McKay, New York.

Gilbert, J. L. 1969. Germinating *Mucuna urens*. Gard. Chron. and New Hort. *166* (2): 5.

Griswold, W. R. 1951. Gulf Stream is yielding its secrets. Sperryscope *12* (4): 16–18.

Gumprecht, T. E. 1854. Zeitschrift dur Allgemeine Erfunde *3*: 1–524 (cf. pp. 409–432). Reimer, Berlin.

Gunn, C. R. 1968. Stranded seeds and fruits from the southeastern shore of Florida. Garden Journal *18* (1): 43–54; cover.

Gunn, C. R. 1969. Seeds of the United States noxious and common weeds in the Convolvulaceae, excluding the genus *Cuscuta*. Assoc. Off. Seed Anal. Proc. *59*: 101–115.

Bibliography

Gunn, C. R. 1972. *Moonflowers, Ipomoea* section *Calonyction*, in temperate North America. Brittonia 24: 150–168.

Gunn, C. R. 1976. *Merremia discoidesperma*. Its taxonomy and capacity of its seeds in ocean drifting. Econ. Bot. In press.

Gunn, C. R. and J. V. Dennis. 1971. Ocean journeys by mangrove seedlings. Shore and Beach 39 (2): 19–22.

Gunn, C. R. and J. V. Dennis. 1972a. An ancient import—sea hearts from the tropics. Yankee Aug.: 168–169.

Gunn, C. R. and J. V. Dennis. 1972b. Stranded tropical seeds and fruits collected from Carolina beaches. Castanea 37: 195–200.

Gunn, C. R. and J. V. Dennis. 1973. Tropical and temperate stranded seeds and fruits from the Gulf of Mexico. Contri. Marine Sci. 17: 111–122.

Gunnerus, J. C. 1765. Efterretninger om de sakaldte Løsnings-stene eller vetter-Myrer, om ormestene og nogle andre udenlandske frugter som findes hist og her ved stranden i Norge. Trondhiemske Selskabs Skrifter 3: 15–32.

Guppy, H. B. 1890. Dispersal of plants as illustrated by the flora of the Keeling or Cocos Islands. Jour. Trans. Victoria Institute 24: 267–306.

Guppy, H. B. 1906. Observations of a naturalist in the Pacific between 1896 and 1899. 2: 1–627. Macmillan, London.

Guppy, H. B. 1917. Plants, seeds, and currents in the West Indies and Azores. 531 pp. Williams and Norgate, London.

Hemsley, W. B. 1885. Reports on botany of Bermuda and various other islands of Atlantic and southern oceans. In: C. W. Thomson and J. Murray, eds. Report on scientific results of voyage of H. M. S. Challenger. Botany 1. cf. pp. 276–313; plates 64–65. Longmans, London.

Hemsley, W. B. 1892. A drift-seed (*Ipomoea tuberosa* L.). Ann. Bot. 6: 369–373.

Heyerdahl, T. 1968. Sea routes to Polynesia. 323 pp. Rand McNally, Chicago.

Heyerdahl, T. and E. N. Ferdon, Jr. 1961. Archaeology of Easter Island. 1: 1–557. Rand McNally, Chicago.

Hobbs, K. 1969. *Entada scandens*. Gard. Chron. & New Hort. 166 (6): 4.

Johnston, I. M. 1949. Botany of San José Island (Gulf of Panama). Sargentia 8: 1–306.

Jonston, J. 1662. Dendrographias sive historiae naturalis de arboribus et fruticibus tam nostri quam peregrini orbis libri decem. 477 pp. Merian, Frankfort.

Kohl, J. G. 1868. Geschichte des Golfstroms und seiner Erforschung. 244 pp. Muller, Bremen.

Kohlmeyer, J. 1968. Marine fungi from the tropics. Mycologia 60: 252–270.

Bibliography

Kohlmeyer, J. and E. Kohlmeyer. 1971. Marine fungi from tropical America and Africa. Mycologia 63: 831–861.

Kotzebue, O. von. 1821. Voyage of discovery into the South Seas and Beering's straits. *3:* 1–422; 2 maps (cf. pp. 155–156). Longman, Hurst, et al., London.

Leenhouts, P. W. 1968. Tropische zaden op de Nederlandse kust. Gorteria *4:* 95–98.

Lindman, C. 1882. Om drifved och andra af hafsströmmar uppkastade naturföremal vid norges kuster. Gotesborgs Kongl. Vetenskaps- och Vitterhets- Samhalles Handlingar *17:* 1–105.

Linnaeus, C. 1789. Rariora norvegiae. Amoenitates Academicae 7: 477–478.

Little, E. L., Jr. and F. H. Wadsworth. 1964. Common trees of Puerto Rico and the Virgin Islands. 548 pp. A H 249, U.S. Dept. Agri., Washington.

Lloydd-Jones, A. 1893. A drift-seed from Swansea Bay. Kew Bull. *1893* (76–77): 114.

Main • lines. 1971. Hollis Engley of Main's. Main • lines Dec. 7.

Martin, M. 1703. A description of the Western Islands of Scotland. In: J. Pinkerton. 1809. A general collection of the best and most interesting voyages and travels in all parts of the world *3:* 572–699. Longman, Hurst, et al., London.

Martins, C. 1857. Experiences sur la persistance de la vitalité des graines flottant a la surface de la mer. Bull. Soc. Bot. France *24:* 324–338.

Mason, R. 1961. Dispersal of tropical seeds by ocean currents. Nature *191:* 408–409.

Morison, S. E. 1942. Admiral of the ocean sea: a life of Christopher Columbus. 680 pp. Little, Brown, Boston.

Morris, D. 1895. A Jamaica drift-fruit. Nature *35:* 64–66.

Morris, D. 1899. A Jamaica drift-fruit. Nature *39:* 322–323.

Muir, J. 1930. *Afzelia bijuga* seeds in the South African beach drift. Jour. Med. Assoc. So. Africa *4:* 355.

Muir, J. 1937. Seed-drift of South Africa. 108 pp. + 15 plates. So. African Depart. Agri. & Forestry. Bot. Sur. Mem. No. 16.

Necker de Saussure, L. A. 1821. Voyage en écosse et aux iles Hébrides *3:* 22–23. Paschoud, Geneva.

Pena, P. and L'Obel, M. de. 1570. Stirpium adversaria nova. (cf. p. 395). Authors, London.

Petiver, J. 1702. Gazophylacii naturae et artis, decas secunda. 128 pp. Author, London.

Petiver, J. 1764. Gazophylacii naturae et artis, decas decima. 62 pp. Author, London.

Bibliography

Prance, G. T. and M. Freitas da Silva. 1973. Caryocaraceae. Flora Neotropica Monograph 12. Hafner, New York.
Pyrard, F. 1611. Discours du voyage des Francais aux Indes Orientales. Paris.
Ridley, H. N. 1930. Dispersal of plants throughout the world. 744 pp. Reeve, London.
Rochon, A. M. de. 1802. Voyages a Madagascar, a Maroc, et aux Indes Orientalis. 2: 145–150. Prault, Paris.
Rudd, V. E. 1967. *Oxyrhynchus* and *Monoplegma* (Leguminosae). Phytologia 15: 289–294.
Safford, W. E. 1905. Useful plants of the island of Guam. Contri. U.S. Nat. Herb. 9: 1–416.
Sauer, J. 1964. Revision of *Canavalia*. Brittonia 16: 106–181.
Schimper, A. F. W. 1891. Indo-malayische strandflora. 205 pp. Fischer, Jena.
Sernander, R. 1901. Den skandinaviska vegetationens spridningsbiologi. 459 pp. Friedlander, Berlin; Lundequistska, Uppsala.
Shackleton, K. 1959. *Entada gigas*. Yachting World, Feb.
Sibbald, R. 1694. Scotia illustrata sive prodromus historiae naturalis Scotiae. Edinburgh.
Sloane, H. 1696. Catalogus plantarum qua in insula Jamaica. 275 pp. Author, London.
Sloane, H. 1707. A voyage to the islands Madera, Barbados, Nieves, S. Christophers, and Jamaica with the natural history of the herbs and trees, four-footed beasts, fishes, birds, insects, reptiles, etc. 1: 1–264. Author, London.
Standley, P. C. 1924. Trees and shrubs of Mexico. Contri. U.S. Nat. Herb. 23 (1–5): 1–1721 (cf. p. 506).
Standley, P. C. and J. A. Steyermark. 1946. Flora of Guatemala, part V. Fieldiana 24 (5): 1–502.
Standley, P. C. and L. C. Williams. 1970. Flora of Guatemala. Fieldiana 24 (9): 1–85.
Ström, H. 1762. Beskrivelse ober fogderiet fondmer. 1: 1–572. Sorge, Copenhagen.
Sykes, W. R. 1970. *Ipomoea pes-caprae* (L.) R. Br. ssp. *brasiliensis* (L.) Ooststr. in the New Zealand botanical region. New Zealand Jour. Bot. 8: 249–253.
Sykes, W. R. and E. J. Godley. 1968. Transoceanic dispersal in *Sophora* and other genera. Nature 218: 495–496.
Treub, M. 1888. Notice sur la nouvelle flore de Krakatau. Ann. Jardin Bot. Biutenzorg 7: 213–223.
Urban, I. 1877. Humiriaceae. In: C. F. P. Martius. Flora brasiliensis 12 (2): 434–454 + plates 92–96. Monach, Leipzig.

Bibliography

van Leeuwen, W. M. D. 1929. Krakatau, part 2. pp. 57–79. Fourth Pacific Science Congress, Batavia.

van Zwaluwenburg, R. H. 1942. Notes on the temporary establishment of insect and plant species on Canton Island. The Hawaiian Planters' Record 46: 49–52.

Vincent, D. M. 1957. Tropical flotsam is profuse on North Island beach these days. Christchurch Star-Sun, April 2, p. 12.

Wallace, J. 1693 and 1700. Descriptions of the isles of Orkney (edited by J. Small, 1883). 251 pp. Brown, Edinburgh.

Warrack, A. 1965. Scots dialect dictionary (cf. p. 113). University of Alabama Press. University.

Watt, J. M. and Breyer-Branddwijk, M. G. 1962. Medicinal and poisonous plants of southern and eastern Africa. 1457 pp. Livingstone, Edinburgh and London.

Williams, L. O. 1973. An antivenin, a pacifier and a bit of botanical sleuthing. Econ. Bot. 27: 147–150.

Index

Acacia thorns, Fig. 3
Acrocomia spp., 50, 172, Figs. 13, 72
Adams, C. Dennis, xi
Africa, 9, 21, 33, 34
African oil palm, 174
Alaska, 40
Aldabra Island, 38
Aleurites J. R. & G. Forst., 4, 50
Aleurites fordii Hemsley, 100
Aleurites moluccana (L.) Willd., 38, 100, Figs. 13, 35
Aleurites sp., 100, Fig. 35
Alligator-apple, 74
Amaryllidaceae, 68
Amazon River, Brazil, 12, 34
Amulet, 17, 22
Anacardiaceae, 70–73
Anacardium occidentale L., 132
Anchovy-pear, 124, Fig. 47
Andira galeottiana Standley, 32, 128, Fig. 48
Andira inermis (W. Wright) H. B. K., 128, Fig. 48
Andrews, Joann, 32
Angel, Rosemary, xi
Angola, 35
Annona glabra L., 74, Fig. 21
Annona squamosa L., 74, Fig. 21
Annona sp., 74, Fig. 21
Annonaceae, 74
Antidote caccoon, 96, Fig. 33
Antidote vine, 13, 96, Figs. 4, 33
Apocynaceae, 76
Asian swamp-lily, 68
Asiatic coast, 37
Astrocaryum mexicana Liebm., 174

Astrocaryum standleyana L. H. Bailey, 174
Astrocaryum spp., 29, 40, 50, 174, Figs. 13, 73
Atlantic Beach, North Carolina, 11, 31
Australia, 36
Avery, George N., xi
Avicennia L., 48
Avicennia germinans (L.) L., 4, 13, 78, Fig. 23
Avicennia nitida Jacq., 78
Avicenniaceae, 78
Ayensu, Edward S., xi
Azores, 29, 56

Bali, 36
Bamboo, 13
Barents Sea, 28
Barra, Hebrides, 86
Barringtonia asiatica (L.) Kurz, 4, 36, 38, 80, Figs. 3, 24
Barringtoniaceae, 80
Bats, 36
Bay-bean, 130, Figs. 13, 50
Beach, lower strand, 42, Fig. 10
Beach, upper strand, 42, Fig. 11
Beachcombing, suggestions for, 42
Beauty-leaf, 110
Bent-stone, 17, 148
Bergen, Norway, 27
Bignoniaceae, 82
Birds, 36, 38
Black-bean tree, 134
Black mangrove, 78
Black palm, 174
Black pearl, 198, Figs. 13, 89
Black River, Jamaica, 33

(233)

Index

Black walnut, 122, Fig. 46
Bloodwood, 13, 166, Fig. 69
Bogue Banks, North Carolina, 11, 31
Bonney, J. R., 50
Borassus flabelliformis L., 176
Borassus spp., 176, Fig. 74
Boulogne, France, 28
Box fruit, 12, 37, 80
Brazil, 34
British Isles, 17
Bruguiera gymnorrhiza Lam., 194, Fig. 88
Buoyancy principles, 4, Fig. 3
Buoyancy, testing, 59
Burning-bean, 21, 158
Burseraceae, 84
Buttons, 21

Cabbagebark, 127, Fig. 48
Caesalpinia L., 4, 8, 15, 50
Caesalpinia bonduc (L.) Roxb., 16, 29, 38, 40, 56, 128, Figs. 5, 8, 13, 49
Caesalpinia major (Medicus) Dandy & Exell, 40, 128, Fig. 49
Caesalpinia sp., 128, Fig. 49
Calabash, 82, 101, Figs. 8, 25
Calatola costaricensis Standley, 118, Fig. 44
Calatola venezuelana Pittier, 118
Calocarpum mammosum (L.) Cronquist, 29, 200, Fig. 90
Calcocarpum spp., 200, Fig. 90
Calophyllum brasiliensis Camb., 110
Calophyllum calaba L., 110, Fig. 40
Calophyllum inophyllum L., 36, 38, 110, Fig. 40
Cameroons, 35
Canarium L., 4
Canarium decumanum Gaertner, 84, Figs. 3, 26
Canarium mehenbethune Gaertner, 84, Fig. 26
Canarium sp., 84, Fig. 26
Canary Islands, 35
Canavalia Adanson, 4, 50, 59
Canavalia bonariensis Lindley, 130
Canavalia cathartica Thours, 130, Fig. 50
Canavalia nitida (Cav.) Piper, 130, Fig. 50
Canavalia rosea (Sw.) DC., 51, 130, Figs. 13, 50
Canavalia sericea A. Gray, 130
Candlenut, 37, 100, Figs. 13, 35
Cannonball tree, 168, Fig. 70
Canton Island, Phoenix Islands, 38
Cape Canaveral, Florida, 31
Cape Charles, Virginia, 31
Cape Cod, Massachusetts, 20, 31

Cape Fear, North Carolina, 31
Cape Hatteras, North Carolina, 70
Cape of Good Hope, South Africa, 9, 35
Carapa guianensis Aublet, 168, Fig. 70
Carapa molluccensis Lam., 168, Fig. 70
Caribbean, 21, 27, 33
Carolina beaches 11, 31, 82
Carya aquatica (Michx. f.) Nutt., 120, Fig. 45
Carya glabra (Mill.) Sweet, 120, Fig. 45
Carya illinoensis (Wang.) K. Koch, 120, Fig. 45
Carya tomentosa Nutt., 120, Fig. 45
Caryocar glabrum (Aublet) Persoon, 86, Fig. 27
Caryocar microcarpum Ducke, 86, Fig. 27
Caryocar villosum (Aublet) Persoon, 29, 86
Caryocaraceae, 86
Cashew, 29, 132
Cassia fistula L., 132, Fig. 51
Cassia grandis L. f, 132, Fig. 51
Castanospermum australe A. Cunn, 134, Fig. 52
Ceiba pentandra (L.) Gaertner, Fig. 3
Cerbera L., 4
Cerbera manghas L., 76, Figs. 3, 22
Cerbera odollam Gaertner, 36, 40, 76, Fig. 22
Chile, 41
Clipperton Island, 40
Clocker, Mrs., collection, Fig. 9
Coco-de-mer, 22, 35, 186, Figs. 6, 7, 81-83
Cocoid palm species, 40, 178, Fig. 75
Coconut, 11, 13, 16, 29, 37, 182, Figs. 78-80, 83
Cocos-Keeling Islands, 12
Cocos nucifera L., 4, 13, 16, 36, 182, Figs. 78-80, 83
Cohune, 178, Fig. 75
Coin plant, 136, Fig. 53
Collecting and uses, 42
Collecting beaches, 27
Columbus bean, 16
Columbus, Christopher, 16
Combretaceae, 88-90
Confinement stone, 17
Congo River, 35
Convolvulaceae, 92-95
Coote, W. B., caption Fig. 7
Coralbean, 13, 148, Figs. 13, 59
Corals, 10
Cork, 4
Cork float, Fig. 2
Cornwall, England, 17, 28
Corozo palm, 172

(234)

Index

Costa Rica, 41, caption Fig. 93
Counter current, Equatorial (Pacific Ocean), 10, 36, 40
  Equatorial (South Atlantic Ocean), 21
Counters in games, 21
Country-almond, 13, 41, 46, 88, Figs. 4, 28
County Cork, Ireland, 200
County Kerry, Ireland, 29
Crabs, land, 9
  robber, 9
Crabwood, 168, Fig. 70
Crescentia cujete L., 29, 82, Figs. 8, 25
Crinum L., 48
Crinum americanum L., 68, Fig. 18
Crinum asiaticum L., 68, Fig. 18
Crospunk, 19
Crucifixion-bean, 19
Cucurbitaceae, 96
Curieuse Island, Seychelles, 22, 186
Current map, xii–xiii
Currents, Agulhas, 9, 34
  Benguela, 35
  Brazil, 34
  California, 40, 88
  Canary, 35
  East Australia, 36
  Equatorial counter (South Atlantic Ocean), 21 (Pacific Ocean), 10, 36, 40
  Gulf Stream 16, 31
  Gulf Stream system, 16, 27
  Humboldt, 40
  Irminger, 28
  Kuroshio, 37, 41, 88
  Monsoon, 36
  Northeast Atlantic, 28
  North Equatorial (Atlantic), 29
  North Equatorial (Pacific), 35, 37, 40, 45, 48, 51
  North Pacific, 41
  Norway, 28
  Portugal, 29
  South Equatorial (Atlantic), 21, 27, 33
  South Equatorial (Pacific), 36, 37, 47
  Yucatan, 32
Cycadaceae, 98
Cycas circinalis L., 98, Fig. 34

Dalbergia ecastaphyllum (L.) Taub., 136, Fig. 53
Dalbergia monetaria L. f., 136, Fig. 53
Darwin, Charles, 4, 56
Degener, Otto and Isa, 38

Delonix regia (Hooker) Raf., 138, Fig. 54
Dendrosicus latifolius (Miller) A. Gentry, 82, Fig. 25
Denmark, 28
Dennis, Eeda Sissener, x
Devon, England, 28
Dioclea Sprengel, 4, 15, 50
Dioclea megacarpa Rolfe, 40, 140, Fig. 55
Dioclea panamensis Walpers, 140
Dioclea reflexa Hooker f., 29, 40, 140, Figs. 3, 5, 8, 13, 14, 55
Dioclea violacea Bentham, 140
Dioclea wilsonii Standley, 140
Dioclea sp., 140, Fig. 55
Djawa, 25, 36
Donegal Bay, Ireland, 29
Double-coconut, 186
Drift fruits, defined, 3
Drift seeds, defined, 3
Driftwood, 10, 28
Duala, Cameroons, 35
Duke, James A., x

Easter Island, 198
Eastern Asia, 37, 40
Ecuador, 12, 41
Edwards, Corinne, E., x, 52
Egg Fruits, 200, Fig. 90
Elaeis guineensis Jacq., 174
Elvis, Fig. 10
England, 17, 28
English Channel, 29
English walnut, 122, Fig. 46
Entada Adanson, 4, 8, 15, 50
Entada formosana Kanehira, 144
Entada gigas (L.) F. & R., 16, 29, 40, 50, 142, Figs. 5, 8, 13, 56
Entada gogo (Blanco) Johnston, 144
Entada koshunensis Hay. & Kanehira, 144
Entada phaseoloides (L.) Merrill, 38, 40, 50, 142, 144, Figs. 3, 57
Entada scandens Bentham, 142, 144
Enterolobium cyclocarpum (Jacq.) Griseb., 146, Fig. 58
Enterolobium timbouva Martius, 146, Fig. 58
Erythrina variegata L. var. *orientalis* (L.) Merrill, 148
Erythrina spp., 4, 12, 36, 48, 50, 148, Figs. 13, 59
Euphorbiaceae, 100–104
Evil eye, 17

Faeroe Islands, 20, 28
Fagaceae, 106
Fat kidney, 17

Index

Favas de Colom, 16
Fern-palm, 98, Fig. 34
Fevernut, 128
Fevillea cordifolia L., 13, 96, Figs. 4, 33
Fiberous balls, Fig. 3
Fiji Islands, 10, 38, Fig. 9
Flacourtiaceae, 108
Flamboyant-tree, 138
Flame-tree, 138
Florida, 20, 31
Florida Keys, 31
Florida Straits, 32
Flotation tests, 59
Form evocative, 25
Fosberg, F. Ray, 38
Fossil-prune, 100
Fowler, Larry, x
France, 28
Franklin, Benjamin, 16
Fungus, terrestrial, 4, Fig. 3
Fungus, tropical marine, 11

Galls, 4, Fig. 3
Garden of Eden, 24
General Gordon, 25
Golden shower, 132, Fig. 51
Golf ball, 190
Golfo de Nicoya, caption Fig. 93
Gourd, 82
Grank Banks of Newfoundland, 20, 28
Grand Cayman Island, 9
Gray nickernut, 16, 21, 28, 29, 34, 37, 128, Figs. 5, 8, 13, 49
Great Barrier Reef, 37
Greenland, 27
Grenada, 33
Grias cauliflora L., 4, 124, Figs. 3, 47
Guam, 38
Guilandina L., 50, 65, 128
Gulf coast of United States, 32
Gulf of Guayaquil, Ecuador, 41
Gulf of Guinea, 13
Gulf of Mexico, 27, 41
Gulf of Panama, 10, 13
Gulf of Siam, 37
Gulf Stream, 16, 31
Gulf Stream system, 16, 27
Gunn, Charles R., Fig. 7
Guppy, H. B., 8
Guittiferae, 110–113

Hawaii or Hawaiian Islands, 10, 40
Hebrides, 17, 86
Heritiera littoralis Dryander, 36, 38, 204, Fig. 92

Hernandia guianensis Aublet, 114
Hernandia nymphiifolia (Presl) Kubitzi, 36, 114, Fig. 42
Hernandia sonora L., 114, Fig. 42
Hernandiaceae, 114
Hippomane mancinella L., 4, 46, 102, Fig. 36
Hippomane sp., 102, Fig. 36
Hog-plum, 13, 72, Figs. 4, 20
Honey mangrove, 78
Horse eye-bean, 158
Hough, Francis Z., 200
Howard, Richard, x
Humiriaceae, 116
Huntington Beach, South Carolina, 31
Hymenaea courbaril L., 31, 57, 150, Fig. 60

Icacinaceae, 118
Iceland, 27
Identification, 46
India, 22, 45
Indian-almond, 88, Fig. 28
Indian Ocean, 9, 22, 35
Indo-Malaysian region, 38
Indonesia, 9, 36
Inner Hebrides Islands, 29
Intsia bijuga (Colebr.) O. Kuntze, 50, 152, Figs. 13, 61
Ipomoea alba L., 92, Fig. 31
Ipomoea macrantha R. & S., 92, Fig. 31
Ipomoea pes-caprae (L.) R. Br., 36, 92, Fig. 31
Ipomoea spp., 4, 48, 92, Fig. 31
Ireland, 19, 28, 56, Fig. 5
Isle of Lewis, Outer Hebrides, 28
Isle of Wight, 29
Isthmus of Panama, 9
Ivory nut palm, 31, 168, Fig. 70

Jamaica, 21, 33
Jamaican navel spurge, 104
Jamaican-walnut, 100
Japan, 37
Java, 25, 36
Jessen, C. B., 53, Fig. 16
Jewelry, sea-bean, 21, 53
Jordaan, P. G., 34
Juglandaceae, 31
Juglans cinerea L., 122, Fig. 46
Juglans jamaicensis C. DC., 122, Fig. 46
Juglans L., 4
Juglans nigra L., 122, Fig. 46
Juglans regia L., 122, Fig. 46
Jutland, Denmark, 28

Index

Kaw Tao Island, 37
Kermadec Islands, New Zealand, 37
Kiem, Stanley, xi
Killigrew, Dame Catherine, 17
Kingston, Jamaica, 33
Kohlmeyer, J. & K. 11
Kohlmeyer, Jan, xi
Krakatua, 36, 80

Lagenaria siceraria (Mol.) Standley, 82
Large ear pod, 146, Fig. 58
Laurelwood, 110
Leaf, largest American spermatophyte, 178
Leaf blade, largest entire, 190
Lecythidaceae, 124
Legume, 56
Leguminosae, 126–167
Leppik, Elmar E., xi
Lesser Sunda Islands, 36
Linnaeus, C., 56
Lodoicea maldivica (Gmel.) Persoon, 22, 186, Figs. 6, 7, 81–83
Lofoten Islands, Norway, 28
Looking-glass tree, 204
Lourenço Marques, Mozambique, 34
Lucuma Molina, 200

Madagascar, 35
Madeira, 36
Mageray, 35
Mahé de la Bourdonnais, 22
Mahogany, 170, Fig. 71
Malaysia, 31, 108
Maldive Islands, 22, 186
Mammea americana L., 112, Fig. 41
Mammee-apple, 112, Fig. 41
Manchineel, 46, 102, Fig. 36
Mangifera indica L., 13, 70, Fig. 19
Mango, 13, 70
Mangrove, 11, 78, 194
Manicaria plukenetii G. & W., 190
Manicaria saccifera Gaertner, 13, 29, 34, 109, Figs. 8, 84
Manning, Wayne E., 120, 122
Map, current, xii–xiii
Marine organisms, Fig. 12
Martha's Vineyard, Massachusetts, 31, 57
Maryland, 31
Mary's-bean, 16, 19, 21, 22, 24, 27, 94, Figs. 5, 8, 13, 32
Mason, Ruth, xi
Massachusetts, 31, 57
Mastichodendron capiri (A. DC.) Cronquist var. *tempisque* (Pitt.) Cronquist, 202, Fig. 91

Mastichodendron foetidissimum (Jacq.) H. J. Lam, 202, Fig. 91
Maximiliana caribaea Griseb. & Wendl., 178, Fig. 76
Maxwell, Richard, 140
Medicines, 21
Meliaceae, 168–171
Melville, A. R., 102
Merremia Dennst., 4, 50
Merremia discoidesperma (Donn. Sm.) O'Donell, 16, 19, 29, 38, 94, Figs. 5, 8, 13, 32
Merremia tuberosa (L.) Rendle, 39, 94, Fig. 32
Miller, Eleanor, Fig. 10
Milk, clearing, 19
Mockernut, 120, Fig. 45
Molucca-bean, 19
Moluccas of East Indies, 19
Monkey apple, 74
Moore, Jr., Harold, 178
Mora excelsa Bentham, 154, 156, Fig. 62
Mora oleifera (Triana) Ducke, 156, Figs. 63, 83
Mora sp., 154, 156, Fig. 63
Mora spp., 48
Moreton Bay chestnut, 134, Fig. 52
Morocco, Fig. 3
Mossman, Robert D., x
Mozambique, 34
Mucuna fawcettii Urban, 4, 158, Fig. 64
Mucuna gigantea (Willd.) DC., 37, 38, 158, Fig. 65
Mucuna mutisiana (HBK) DC., 40
Mucuna nigricans Steudel, 158, Fig. 65
Mucuna sloanei F. & R., 4, 16, 29, 40, 158, Fig. 64
Mucuna urens (L.) Medicus, 4, 10, 16, 29, 40, 50, 158, Fig. 65
Mucuna spp., 8, 15, 17, 48, 50, 158, Figs. 5, 8, 13, 14
Muir, John, 21, 34
Mull, Inner Hebrides, 29
Myrtle Beach, South Carolina, 31

Nantucket, Massachusetts, 38
Narvik, Norway, 28
Natal, Brazil, 34
Netherlands, 28
New Guinea, 36
New World, western, 40
New Zealand, 36, 137
Ninety Mile Beach, New Zealand, 47, 92, 137
North Atlantic, 16, 27
North Cape, Mageray, 28
North Carolina, 31, 70

Index

North Island, New Zealand, 37
North Sea, 29
North Uist, Outer Hebrides, 28
Northeast Passage, 19
Northern Europe, 17, 27
Norway, 17, 27
Nova Zembla, 28
Nypa, 192, Fig. 85
Nypa fruticans Wurmb., 192, Fig. 85

Oahu, Hawaii, 38
Oaks, 106, Fig. 38
Ocean City, Maryland, 31, 200
Omphalea diandra L., 4, 104, Fig. 3
Omphalea panamensis (Beurl.) I. M. Johnston, 104
Omphalea triandra L., 104, Fig. 3
Orbignya cohune (Mart.) Standley, 178, Fig. 77
Oregon coast, 41, 88
Orinoco River, Brazil, 34
Orkneys, 19, 28
Outer Hebrides, 28
Oxyrhynchus T. S. Brandegee, 50
Oxyrhynchus trinervius (Donn. Sm.) Rudd, 162, Figs. 13, 67
Oxyrhynchus volubilis Brandegee, 162

Pacific Ocean, 38
Padre Island, Texas, 32
Palisadoes Beach, Jamaica, 33
Palmae, 4, 172–193
Palm Beach, Florida, 32
Pandanus spp., 38
Pangium edule Reinw., 108, Fig. 39
Parapara, 198
Parks, Pat, x
Passiflora spp., 82
Passionflower, 82
Pawleys Island, South Carolina, 31
Peacock-tree, 138
Pecan, 120, Fig. 45
Pelliciera rhizophorae Planch & Triana, 41, 206, Fig. 93
Peltophorum inerme (Roxb.) Naves, 4, 10, 164, Figs. 3, 68
Philippines, 37
Phoenix Islands, 38
Phytelephas macrocarpa Ruiz. & Pav., 31, 168, Fig. 70
Pico, 31
Pignut, 120, Fig. 45
Pink shower, 132, Fig. 51
Piscidia erythrina L., 148
Pitcairn, 10
Pointe Noire, Congo, 35
Polynesia, 10

Ponce de Leon, 20
Pond apple, 74
Porto Santo, Azores, 16
Portsmouth, England, 29
Portugal, 29
Pouteria Aublet, 200
Praslin, Seychelles, 22, 186
Prickly palm, 172, Figs. 13, 72
Proctor, George, xi
Progreso, Yucatan, 32
Pterocarpus officinalis Jacq., 13, 166, Fig. 69
Pterocarpus sp., 166, Fig. 69
Puerto Rico Trench, 13, caption Fig. 4
Pulu We, 36
Pumice, 4, 10, 13, Fig. 2
Puntarenas, Costa Rica, 41
Puzzle fruit, 204, Fig. 92

Queen sago, 98
Quercus bennettii Miq., 106, Fig. 38
Quercus spp., 106, Fig. 38
Quintana Roo, 32

Railroadvine, 92
Ratak Chain, 38
Recife, Brazil, 34
Red mangrove, 11, 12, 13, 194, Figs. 4, 86, 87
Rhizomes, 10
Rhizophoraceae, 194
Rhizophora mangle L., 10, 13, 194, Figs. 4, 86, 87
Rhizophora mucronulata Lam., 194, Fig. 88
Rhizophora samoensis (Hochr.) Salvoza, 4, 194
Rich, Jerry, Fig. 15
Rosenstiel School, 13
Royal poinciana, 138
Ryukyu Islands, 37

Sacoglottis amazonica Martius, 4, 29, 116, Figs. 3, 8, 43
Saddle-bean, 140
St. Croix, 33
Salt-bush, 78
San Jose Island, 10, 40
San Miguel, Azores, 31
Santa Rosa Island, Florida, 32
Sapindaceae, 198
Sapindus L., 50
Sapindus saponaria L., 31, 40, 198, Figs. 13, 89
Sapotaceae, 200–203
Scaevola koenigii Vahl, 36

(238)

Index

Scotland, 29
    amulet, 17, 22
Sea-bean, buoyancy, 4, 59, Fig. 1
    catalog, 61–207
    charms, 16, 21
    cleaning, 50
    collecting, 42
    collecting beaches, 27–41
    defined, 3
    dredging, 13, Fig. 4
    economic aspects, 11
    flotation test, 59
    foreign collecting, 48
    germinating, 56
    growing, 56
    history, 16
    identification, 46, 61
    jewelry, 49, 53, Fig. 17
    key, 61
    lists, 211, 217
    literature, 229
    observations at sea, 12
    passing though customs, 48
    polishing, 59, Figs. 13, 14, 16, 17
    preserving fleshy species, 48
    scarification, 57
    sectioning, 46
    storing, 48
    transport currents, xii, 27–41
    uses 16, 42
    viability, 56, 60
    world lists, 211, 217
Sea-coconut, New World, 12, 13, 34, 190, Figs. 8, 84
Sea-coconut, Old World, 186
Sea heart, 16, 17, 21, 28, 29, 34, 37, 56, 57, 142, Figs. 5, 8, 13, 56
Sea moonflower, 92
Sea pencil, 194
Sea purse, 16, 17, 21, 29, 34, 57, 140, Figs. 5, 8, 13, 55
Seed embryo, largest, 156
Seychelles, 22, 34, 186
Shetland Islands, 19, 28
Shipworms, 11
Singapore, 37
Skye, Inner Hebrides, 29
Sleeve palm, 190
Sloane, Sir Hans, 20
Small ear pod, 146, Fig. 58
Smith, Jr., Earle, x, 74
Smithsonian Institution, xi
Snuffbox sea-bean, 144, Fig. 57
Snuffboxes, 17, 21
Soapberry, 198
Solomon Islands, 12
Somalia Republic, 28

Sophora tomentosa L., 10
South Africa, 28, 70
South America, eastern, 33
South America, western, 40
South Carolina, 31
South Devon, England, 200
Southern swamp lily, 68
South Uist, England, 28
Spermatophytes, 4
Spice Islands of East Indies, 19
Spitzbergen, 28
Spondias dulcis S. Parkinson, 72, Fig. 20
Spondias mombin L., 13, 72, Fig. 20
Spondias purpurea L., 72
Starnut palm, 174, Figs. 13, 73
Stems, 10
Sterculiaceae, 170
Strongylodon lucidus (Forst. f.) Seem., 40, 162, Fig. 67
Strongylodon Vogel, 50
Sugar-apple, 74
Sugar cane, stalks of, 13
Sumatera, 25, 36
Sumatra, 25, 36
Swamp-lily, 68
Sweet pignut, 120, Fig. 45
Sweet sop, 74
Swietenia mahagoni (L.) Jacq., 170, Fig. 71
Sydney, 37
Systematic descriptions, 58
Systematic illustrations, 58

Tahiti, 10
Taiwan, 37
Teething ring, 17
Teredos, 11, 166
Terminalia catappa L., 13, 36, 41, 46, 88, Figs. 4, 28
Terminalia spp., 4, 88, Figs. 29, 30
Tetrazolium, 60
Texas, 32
Theaceae, 206
Thorns, 4
Thorns, *Acacia*, Fig. 3
    Ceiba pentandra (L.) Gaertner, Fig. 3
Tieghemella Balle, 200
Tierra del Fuego, 41
Tobago, 33
Toys, 32
Transport currents, xiii, 27
Tree of Knowledge of Good and Evil, 25
Trinidad, 33
Tristan da Cunha, 140
Tromsoe, Norway, 28
Tropical-almond, 88
Tropical drift seeds and fruits defined, 3

Index

True sea-bean, 16, 17, 23, 28, 29, 34, 57, 142, Figs. 5, 8, 13, 64, 65
Tsunami, 38
Tung, 100
Turks Islands, 33

United States, 16, 20, 31, 40
United States, eastern, 20, 31
United States, western, 40

Valley of the Giants, Seychelles, 25
Vantanea guianensis Aublet, 116, Fig. 43
Veracruz, 32
Viability, testing, 60
Viable seeds, 56
Virginia, 31
Viti Levu, Fiji Islands, 38, Fig. 9
Voss, Gilbert, xi

Water hickory, 120, Fig. 45
West Frisian Islands, Netherlands, 28

West Indian locust tree, 31, 57, 150, Fig. 60
West Indies, 9, 21, 33
White moonflower, 92
White walnut, 122
Wind, 36, 45
Wings Neck, Massachusetts, 31
Wood, 11
Wood-rose, 94
Work, Robert, xi
Wormstone, 17
Wotho Atoll, Marshall Islands, 38

Xylocarpus moluccensis (Lam.) Roem., 168, Fig. 70

Yellow flamboyant, 164
Yellow nickernut, 128
Yucatan, Mexico, 32
Yucatan Channel, 32

Zambezi River, Mozambique, 34